Communications
in Computer and Information Science   **1015**

Commenced Publication in 2007
Founding and Former Series Editors:
Phoebe Chen, Alfredo Cuzzocrea, Xiaoyong Du, Orhun Kara, Ting Liu,
Krishna M. Sivalingam, Dominik Ślęzak, Takashi Washio, and Xiaokang Yang

More information about this series at http://www.springer.com/series/7899

Jong-Hwan Kim · Hyung Myung ·
Seung-Mok Lee (Eds.)

Robot Intelligence Technology and Applications

6th International Conference, RiTA 2018
Kuala Lumpur, Malaysia, December 16–18, 2018
Revised Selected Papers

 Springer

Editors
Editors
Jong-Hwan Kim
Korea Advanced Institute of Science
and Technology (KAIST)
Daejeon, Korea (Republic of)

Hyung Myung
Korea Advanced Institute of Science
and Technology (KAIST)
Daejeon, Korea (Republic of)

Seung-Mok Lee
Keimyung University
Daegu, Korea (Republic of)

ISSN 1865-0929   ISSN 1865-0937 (electronic)
Communications in Computer and Information Science
ISBN 978-981-13-7779-2   ISBN 978-981-13-7780-8 (eBook)
https://doi.org/10.1007/978-981-13-7780-8

This Springer imprint is published by the registered company Springer Nature Singapore Pte Ltd
The registered company address is: 152 Beach Road, #21-01/04 Gateway East, Singapore 189721, Singapore

Preface

The 6th edition of the International Conference on Robot Intelligence Technology and Applications (RiTA 2018) was held in Putrajaya, Malaysia, during December 16–18, 2018. The event was organized by the Korea Advanced Institute of Science and Technology (KAIST) and the Innovative Manufacturing, Mechatronics and Sports (iMAMS) Laboratory, Universiti Malaysia Pahang (UMP). RiTA aims at serving researchers and practitioners in related fields with timely dissemination of the recent progress on robot intelligence technology and its application.

The sixth edition of the conference had the theme of "Robotics and Machine Intelligence: Building Blocks for Industry 4.0." The Fourth Industrial Revolution (IR 4.0) brought technological disruption, unlike its predecessors. Just as innovations in manufacturing processes and systems led previous industrial revolutions, the advancement of IR 4.0 will be driven by a smart, interconnected, and pervasive environment with data being one of its main currencies as well as robotics and automation being the central pillar of its growth.

RiTA 2018 received 80 submissions from eight different countries with mainly from South Korea and Malaysia. All submissions were reviewed in a single-blinded manner, and the best 20 papers recommended by the reviewers are published in this volume. The editors would like to thank all the authors who submitted their work, as the papers are of good quality and represented good progress in robot intelligence and its applications especially in facing the fourth industrial revolution.

The editors would also like to thank Prof. Junmo Kim for delivering his plenary speech entitled "Recent Advances and Challenges in Deep Learning for Computer Vision" as well as Associate Professor Marcelo H. Ang Jr. (National University Singapore, Singapore), Prof. Etienne Burdet (Imperial College London, UK), Prof. Elmer P. Dadios, Assistant Professor João Sequeira (Technical University of Lisbon, Portugal), Dr. Esyin Chew (Cardiff Metropolitan University, UK), Associate Professor Dr. Hanafiah Yussof (Universiti Teknologi MARA, Malaysia) for delivering their speeches at the conference.

The editors hope that readers find this volume informative. We thank Springer for undertaking the publication of this volume. We also would like to thank the conference Organizing Committee for their hard work in realizing the conference.

March 2019

Jong-Hwan Kim
Hyung Myung
Seung-Mok Lee

Organization

Honorary General Chair

Jong-Hwan Kim   KAIST, South Korea

General Chairs

Zahari Taha   UMP, Malaysia
Hyun Myung   KAIST, South Korea

Program Chairs

Dato' Sri Daing Nasir Ibrahim   UMP, Malaysia
Rizalman Mamat   UMP, Malaysia
Han-Lim Choi   KAIST, South Korea

Organizing Chairs

Ahmad Shahrizan Abdul Ghani   UMP, Malaysia
Muhammad Aizzat Zakaria   UMP, Malaysia
Ahmad Fakhri Ab. Nasir   UMP, Malaysia
Junmo Kim   KAIST, South Korea
Weiliang Xu   University of Auckland, New Zealand
Eric T. Matson   Purdue University, USA
Jun Jo   Griffith University, Australia
Fakhri Karray   University of Waterloo, Canada

Special Session Chairs

Ahmad Najmuddin Ibrahim   UMP, Malaysia
Gourab Sen Gupta   Massey University, New Zealand

Workshop/Tutorial Chair

Taesup Moon   Sungkyunkwan University, South Korea

Plenary Session Chair

Ah Hwee Tan   Nanyang Technological University, Singapore

Awards Committee Chair

Serin Lee   KYRA, USA

Local Arrangements Chairs

Rabiu Muazu Musa   UMP, Malaysia
Mohd Fadzil Abdul Rahim   UMP, Malaysia
Ash Choong Chun Sern   UMP, Malaysia
Dong Eui Chang   KAIST, South Korea

Publications Chairs

Anwar P. P. Abdul Majeed   UMP, Malaysia
Mohd Hasnun Arif Hassan   UMP, Malaysia
Jessnor Arif Mat Jizat   UMP, Malaysia
Jin-Woo Jung   Dongguk University, South Korea
Seungmok Lee   Keymyung University, South Korea

Exhibition Chairs

Nafrizuan Mat Yahya   UMP, Malaysia
Zubair Khalil   UMP, Malaysia
Nurul Akmal Che Lah   UMP, Malaysia
Donghwa Lee   Daegu University, South Korea
Donghan Kim   Kyunghee University, South Korea

Secretariat

Nurul Qastalani Radzuan   UMP, Malaysia
Nur Fahriza Mohd Ali   UMP, Malaysia
A. J. Ahn   MSREP, KAIST, South Korea

Local Program Committee

Shahrul Na'im Sidek   IIUM, Malaysia
Zulkifli Mohamed   UiTM, Malaysia
Mohd Zaid Abdullah   USM, Malaysia
Musa Mailah   UTM, Malaysia
Noor Azuan Abu Osman   UMT, Malaysia
Ali Yeon Md Shakaff   UNIMAP, Malaysia
Kamarul Hawari Ghazali   UMP, Malaysia
Ponnambalam Sivalinga   UMP, Malaysia
  Govinda Rajan
Zamberi Jamaludin   UTeM, Malaysia
Mohd Rizal Arshad   USM, Malaysia

Muhammad Azmi Ayub	UITM, Malaysia
Ishkandar Baharin	MyRAS, Malaysia
Rosziati ti Ibrahim	UTHM, Malaysia
Zainal Alimuddin Zainal Alauddinn	USM, Malaysia

International Program Committee

Abdelghani Chibani	Paris Est Creteil University, France
Alan Wee-Chung Liew	Griffith University, Australia
Anna Friesel	Technical University of Denmark, Denmark
Azizi Abdullah	UKM, Malaysia
Bela Stantic	Griffith University, Australia
Brijesh Verma	CQUniversity, Australia
Bumjoo Lee	Myung-Ji University, South Korea
David Claveau	California State University Channel Islands, USA
Donald G. Bailey	Massey University, New Zealand
Guangming Xie	Peking University, China
Hongliang Ren	NUS, Singapore
Hyun-Taek Choi	KRISO, South Korea
Igor Verner	Technion – Israel Institute of Technology, Israel
Julia Taylor	Purdue University, USA
Joao Sequeira	ISR – Instituto Superior Técnico, Portugal
Kaori Yoshida	Kyushu Institute of Technology, Japan
Kojiro Iizuka	Shibaura Institute of Technology, Japan
Maki K. Habib	The American University in Cairo, Egypt
Marley Maria B. R. Vellasco	Pontifical Catholic University of Rio de Janeiro, Brazil
Meng Cheng Lau	University of Manitoba, Canada
Muhammad Hafidz Fazli Md Fauadi	UTeM, Malaysia
Nak Yong Ko	Chosun University, South Korea
Nak-Yong Chong	Japan Advanced Institute of Science and Technology, Japan
Pei-Chen Sun	National Kaohsiung Normal University, Taiwan
Peter Sincak	TU Kosice, Slovakia
Pitoyo Hartono	Chukyo University, Japan
Rini Akmeliawati	IIUM, Malaysia
Rituparna Datta	IIT Kanpur, India
Sang Wan Lee	KAIST, South Korea
Seul Jung	Chungnam National University, South Korea
Sung-eui Yoon	KAIST, South Korea
Sunglok Choi	ETRI, South Korea
Tangwen Yang	Beijing Jiaotong University, China
Wan Chul Yoon	KAIST, South Korea

XiaoPing Chen   University of Science and Technology of China, China

Yuchi Ming   Huazhong University of Science and Technology, China

Contents

About the Editors

Jong-Hwan Kim (FIEEE 2009) received his PhD degree in electronic engineering from Seoul National University, South Korea, in 1987. Since 1988, he has been with the School of Electrical Engineering, KAIST, South Korea, where he leads the Robot Intelligence Technology Laboratory as professor. Dr. Kim is Director for both the KoYoung-KAIST AI Joint Research Center and the Machine Intelligence and Robotics Multi-Sponsored Research and Education Platform. His research interests include intelligence technology, machine intelligence learning, and AI robots. He has authored five books and five edited books, two journal special issues, and around 400 refereed papers in technical journals and conference proceedings.

Hyun Myung received his BS, MS, and PhD degrees from the Korea Advanced Institute of Science and Technology (KAIST), Daejeon, South Korea, in 1992, 1994, and 1998, respectively, all in electrical engineering. He was a senior researcher with the Electronics and Telecommunications Research Institute, Daejeon, from 1998 to 2002; CTO and Director of the Digital Contents Research Laboratory, Emersys Corporation, Daejeon, from 2002 to 2003; and a principal researcher with the Samsung Advanced Institute of Technology, Yongin, South Korea, from 2003 to 2008. From 2008 to 2018, he was a professor with the Department of Civil and Environmental Engineering, KAIST, where he is currently a professor with the School of Electrical Engineering and the head of the KAIST Robotics Program. His current research interests include structural health monitoring using robotics, artificial intelligence, simultaneous localization and mapping, robot navigation, machine learning, deep learning, and swarm robots.

Seung-Mok Lee received his PhD degree in civil and environmental engineering (Robotics Program) from the Korea Advanced Institute of Science and Technology (KAIST), Daejeon, South Korea, in 2014. He was a post-doctoral fellow with the Urban Robotics Laboratory, KAIST, Daejeon, from 2014 to 2015, a senior research engineer with the Intelligent Safety Technology Center, Hyundai Motor Company, Hwaseong, South Korea, from 2015 to 2017. Since 2017, he has been Assistant Professor with the Department of Mechanical and Automotive Engineering, Keimyung University, Daegu, South Korea. His current research interests include soft computing, autonomous vehicles, simultaneous localization and mapping, and robot navigation.

Data-Driven Neuroendocrine-PID Tuning Based on Safe Experimentation Dynamics for Control of TITO Coupled Tank System with Stochastic Input Delay

Mohd Riduwan Ghazali[✉], Mohd Ashraf Ahmad, and Raja Mohd Taufika Raja Ismail

Instrumentation and Control Engineering (ICE) Research Cluster, Faculty of Electrical and Electronics Engineering, Universiti Malaysia Pahang, 26600 Pekan, Pahang, Malaysia
riduwan@ump.edu.my

Abstract. This paper addresses a data-driven neuroendocrine-PID tuning for control a two-input-two-output (TITO) coupled tank system with stochastic input time delay based on safe experimentation dynamics (SED). The SED algorithm is an optimization method used as data-driven tools to find the optimal control parameters by using the input-output (I/O) data measurement in an actual system. The advantages of the SED algorithm are that provides a fast solution, able to solve the high dimensional problem and provides high-performance accuracy by keeping the best parameter value while finding the control parameters. Moreover, the gain sequences of the SED algorithm is independent of the number of iterations by fixed the interval size in finding the optimal solution. Hence, this allows the SED method to have enough strength to re-tune in the attempted of finding the new optimal solution when the delay occurs during the tuning process. Apart from that, a neuroendocrine-PID controller structure is chosen due to its provide effective and accurate control performances by a combination of PID and neuroendocrine structures. On another note, the neuroendocrine structure is a biologically inspired designed that derived from general secretion rules of the hormone in the human body. In order to evaluate the performances of the data-driven neuroendocrine-PID control based on SED, it is applied to a numerical example of TITO coupled tank plant and the control performance tracking and the computational time are observed. The simulation results show that the data-driven neuroendocrine-PID control based on SED capable to track the desired value of liquid tanks level although the stochastic input delay occurred in the system. In addition, the SED based method also attained good control performance without any theoretical assumptions about plant modelling.

Keywords: TITO · Neuroendocrine-PID · Coupled tank · SED · Stochastic delay

J.-H. Kim et al. (Eds.): RiTA 2018, CCIS 1015, pp. 1–12, 2019.
https://doi.org/10.1007/978-981-13-7780-8_1

1 Introduction

Nowadays, a coupled tank system is widely used in industrial sectors especially in the process control, food processing and water purification systems that require accurate liquid level control of storage tanks [1]. Besides, most of the coupled tank system is a type of two-input-two-output (TITO) system [2, 3]. This TITO coupled tank system is a popular system for the very challenging problem of the controller design due to the dynamic process and existence of interactions between input and output variables. In addition, the delays always exist caused by communication or transportation delay and duration to sense the parameter by the sensor for the actual system [4]. For example, the communication delay such as the stochastic input time delay occurred during the process of data transfer from controller to input system in a random manner [4, 5]. Therefore, this input delay could increase the difficulty of the control design and affect the stability of the controller.

So far, various control strategies have been widely reported in order to control the liquid level of coupled tank systems. For example, proportional-integral (PI) with decoupling controller has been reported to control the TITO coupled tank [2]. This control method which is a model-based approach utilized the root locus approach in determining the controller parameters. This is a model-based method that may contribute to several problems, such as the inaccuracy of the simplified model and the huge gap between control theory and real applications. Besides, some researchers proposed the PI control design by using model reference adaptive control (MRAC) to find the optimal controller parameters [3, 6]. This approach is an offline based experiment tuning method such that the tuning process happens without running the actual plant [7]. Normally, offline based experiment process refers to the process of computing the optimal control parameters from the one-shot or two-shot experiment data. Then, these computed control parameters are implemented in the actual system [8]. However, the computed parameters must be re-tuned if the plant structure changes due to disturbance, uncertainties and time delay occurred during the tuning process. Another problem is, not all data sets contain sufficient dynamic information that affected the optimal control parameters obtained. Therefore, the online-based experiment is more effective to determine optimal control parameters by running an actual plant repeatedly in specific time duration until the best control objective is attained [9]. In addition, others controller and control strategy that has been applied to coupled tank system are sliding mode control [10], PI-neural network based on multi-agent search algorithm [1], PID based on PSO [11] and etc.

Here, the online data based experiment which is data-driven control by using pre-specifics fixed PID controller structure is widely used due to its robustness, less complexity in the design and easy to implement [9, 12]. But, this pre-specific fixed PID structure might contribute poor control performances, especially in non-linear, complex system and delays occurred. Therefore, the neuroendocrine-PID controller structure as an enhancement version of the PID structure is designed to solve the non-linear and complex control system. This neuroendocrine-PID is designed based on the secretion of the hormone regulation principle provided better control performance, effective and efficient control design to improve the capability of PID [13, 14].

This paper presents a data-driven neuroendocrine-PID control for TITO coupled tank system with stochastic input delay based on a game-theoretic approach called safe experimentation dynamics (SED) based method. This is due to the SED method produces stable convergence and high accuracy with memorable procedures characteristics [15]. Furthermore, the gain sequence of the SED is independent by fixed the interval size that has enough energy to find the optimal solution when delays occur during the tuning process. Thus, it would be useful to observe the SED based method capability in tuning the neuroendocrine-PID controller for TITO coupled tank system with stochastic input delay. The performances the SED based method is verified in term of the tracking error, the control accuracy and the computational time. The contributions of this work are the SED based method able to track the desired trajectories and provides good accuracy when dealing with stochastic delay since it is memory-based optimization and independent gain sequence in the updated procedure.

The structural organization of this paper as follows. In Sect. 2, the problem formulation or the derivation of neuroendocrine-PID controller structure design in minimizing the error of the TITO coupled tank system. Section 3, represent the SED algorithm procedures while Sect. 4 shows the design of data-driven neuroendocrine-PID control. In Sect. 5, the implementation of the SED approach which is used to find the parameter of neuroendocrine-PID is summarized. Section 6 presents the control performances of results with the discussion. Then, in Sect. 7, a conclusion of findings is presented.

Notation: The symbols $\mathbb{R}$, $\mathbb{R}^n$ and $\mathbb{R}_+$ are denotes as a set of real number, set of n real number and set of positive real number, respectively.

2 Problem Formulation of Neuroendocrine-PID Control

Let consider the TITO coupled tank system for the neuroendocrine-PID controller as illustrated in Fig. 1 where, $u(t) \in \mathbb{R}^2$, $r(t) \in \mathbb{R}^2$ and $y(t) \in \mathbb{R}^2$ are the input control, the reference and the output of the system, respectively. The plant is a TITO coupled tank system denoted as $G(s)$. Then, the time is $t = 0$, $t_s, 2t_s, 3t_s \ldots \ldots Mt_s$ where t_s and M are referred to sampling time and number of samples, respectively. The stochastic input delay is $u_i(t - d)(i = 1, 2) \in \mathbb{R}^2$, where d is random values that generated uniformly between 1 and 3.

The controller $C(s)$ is a PID controller structure is defined as

$$C(s) = \begin{bmatrix} P_{11}\left(1 + \frac{1}{I_{11}s} + \frac{D_{11}s}{1+(D_{11}/N_{11})s}\right) & 0 \\ 0 & P_{22}\left(1 + \frac{1}{I_{12}s} + \frac{D_{12}s}{1+(D_{12}/N_{12})s}\right) \end{bmatrix} \quad (1)$$

where P_{11}, $P_{22} \in \mathbb{R}$, I_{11}, $I_{22} \in \mathbb{R}$, D_{11}, $D_{22} \in \mathbb{R}$ and N_{11}, $N_{22} \in \mathbb{R}$ are refers to proportional gain, integral gain, derivative gain and filter coefficient, respectively. Here, the output of the controller $C(s)$ is denoted as

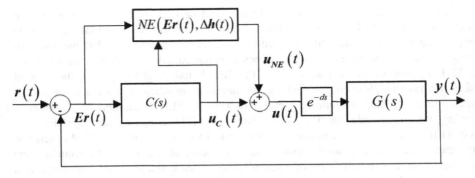

Fig. 1. Neuroendocrine-PID controller with TITO plant with stochastic input delay

$$u_C(t) = [C(s)] \begin{bmatrix} Er_1(t) \\ Er_2(t) \end{bmatrix}. \tag{2}$$

Then, the $NE(\mathbf{Er}(t), \Delta \mathbf{h}(t))$ is defined as

$$NE(\mathbf{Er}(t), \Delta \mathbf{h}(t)) = \begin{bmatrix} E_{11}(Er_1(t), \Delta h_{11}(t)) & 0 \\ 0 & E_{22}(Er_2(t), \Delta h_{22}(t)) \end{bmatrix}. \tag{3}$$

The hormone regulation of the endocrine system follows the hill function [14] is denoted as

$$E_{ij}(Er_j(t), \Delta h_{ij}(t)) = \alpha_{ij} \left[\frac{(|\Delta h_{ij}(t)|)^{\xi_{ij}}}{\lambda_{ij} + (|\Delta h_{ij}(t)|)^{\xi_{ij}}} + \beta_{ij} \right] \cdot \left(-\frac{Er_j(t)}{|Er_j(t)|} \cdot \frac{\Delta Er_j(t)}{|\Delta Er_j(t)|'} \right)$$
$$\cdot \left(\frac{\Delta h_{ij}(t)}{|\Delta h_{ij}(t)|} \right) \tag{4}$$

where the $\Delta h_{ij}(t) = h_{ij}(t) - h_{ij}(t - t_s)$, $\Delta Er_j(t) = Er_j(t) - Er_j(t - t_s)$ and $\alpha_{ij}, \beta_{ij}, \lambda_{ij}, \xi_{ij}$ are variance, change of error and neuroendocrine parameters, respectively. Notes that, $E_{ij}(Er_j(t), \Delta h_{ij}(t)) = 0$ should be satisfied if $\Delta h_{ij}(t) = 0$, thus all β_{ij} should be 0 [14]. Next, the output of the neuroendocrine structure $NE(\mathbf{Er}(t), \Delta \mathbf{h}(t))$ is

$$\mathbf{u}_{NE}(t) = [E_{11}(Er_1(t), \Delta h_{11}(t)) \quad E_{22}(Er_2(t), \Delta h_{22}(t))]^T \tag{5}$$

and the neuroendocrine-PID output is represented as

$$\mathbf{u}(t) = \mathbf{u}_C(t) + \mathbf{u}_{NE}(t). \tag{6}$$

The corresponding control performance according to the control system in Fig. 1 is denoted as

$$\hat{e}_j = \int_{t_0}^{t_f} \left| r_j(t) - y_j(t) \right|^2 dt \tag{7}$$

where $j = 1, 2$. The time interval of performance evaluation is $[t_0, t_f]$ where $t_0 \in \{0\} \cup R_+$ and $t_f \in R_+$. An objective function of neuroendocrine-PID parameters is defined as

$$J(\boldsymbol{P}, \boldsymbol{I}, \boldsymbol{D}, \boldsymbol{N}, \boldsymbol{\alpha}, \boldsymbol{\xi}, \boldsymbol{\lambda}) = w_1 \hat{e}_1 + w_2 \hat{e}_2 \tag{8}$$

where $\boldsymbol{P} := [P_{11} \quad P_{22}]^T$, $\boldsymbol{I} := [I_{11} \quad I_{22}]^T$, $\boldsymbol{D} := [D_{11} \quad D_{22}]^T$, $\boldsymbol{N} := [N_{11} \quad N_{22}]^T$, $\boldsymbol{\alpha} := [\alpha_{11} \quad \alpha_{22}]^T$, $\boldsymbol{\xi} := [\xi_{11} \quad \xi_{22}]^T$, $\boldsymbol{\lambda} := [\lambda_{11} \quad \lambda_{22}]^T$ and the weighting coefficients is denoted as $w_j \in \mathbb{R}(j = 1, 2)$.

Problem 2.1. Find a neuroendocrine-PID controller $C(s)$ and $NE(\boldsymbol{Er}(t), \Delta \boldsymbol{h}(t))$ parameters, which to minimizes the objective function $J(\boldsymbol{P}, \boldsymbol{I}, \boldsymbol{D}, \boldsymbol{N}, \boldsymbol{\alpha}, \boldsymbol{\xi}, \boldsymbol{\lambda})$ with respect to parameter design $\boldsymbol{P}, \boldsymbol{I}, \boldsymbol{D}, \boldsymbol{N}, \boldsymbol{\alpha}, \boldsymbol{\xi}$ and $\boldsymbol{\lambda}$ based on input and output measurement data $(\boldsymbol{u}(t), \boldsymbol{y}(t))$ from the system in Fig. 1.

3 Safe Experimentation Dynamics Approach

Here, let consider

$$\min_{p \in \mathbb{R}^n} f(\boldsymbol{p}) \tag{9}$$

is an optimization problem, where $f : \mathbb{R}^n \to \mathbb{R}$ is an objective function by selection of design parameter $\boldsymbol{p} \in \mathbb{R}^n$. An optimal parameter solution $\boldsymbol{p}_{opt} \in \mathbb{R}^n$ attained by the SED algorithm. Then, the updated law of the SED algorithm is

$$p_i(k+1) = \begin{cases} x(\bar{p}_i - K_g rv_2) & \text{if} \quad rv_1 \leq ET, \\ \bar{p}_i & \text{if} \quad rv_1 > ET, \end{cases} \tag{10}$$

where $k = 0, 1 \ldots k_{\max}$ is number of iterations. The design parameter $p_i \in \mathbb{R}$ is the ith element of $\boldsymbol{p} \in \mathbb{R}^n$ and $\bar{p}_i \in \mathbb{R}$ is the ith element of $\bar{\boldsymbol{p}} \in \mathbb{R}^n$. The $\bar{\boldsymbol{p}}$ is used to keep the present best design parameters value and symbol $rv_1 \in \mathbb{R}$ is the random number which is generated uniformly between 0 and 1. The symbol $rv_2 \in \mathbb{R}$ is a random number and ET is a scalar that defines the probability to use a new random setting for $\boldsymbol{p}$ while K_g is a scalar value representing the interval size to adopt on the random steps on $p_i \in \mathbb{R}$. Next, the selection rule of function x in Eq. (10) as followed:

$$x(\bullet) = \begin{cases} p_{\max} & \bar{p}_i - K_g rv_2 > p_{\max}, \\ \bar{p}_i - K_g rv_2 & p_{\min} \leq \bar{p}_i - K_g rv_2 \leq p_{\max}, \\ p_{\min} & \bar{p}_i - K_g rv_2 < p_{\min}, \end{cases} \tag{11}$$

where $p_{\max}$ and $p_{\min}$ are minimum and maximum defined values of the design parameter. The SED algorithm has several steps to follows as illustrated in Fig. 2.

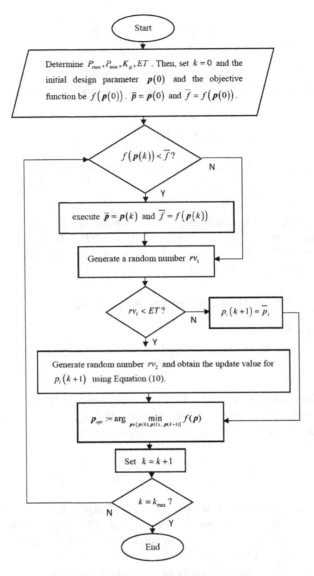

Fig. 2. Flowchart of SED algorithm

4   Data-Driven Neuroendocrine-PID Control Design

In this section, the SED algorithm is implemented for data-driven neuroendocrine-PID control design for TITO coupled tank system with stochastic input delay. Here, the parameters design as follow:

$$\psi = [P, I, D, N, \alpha, \xi, \lambda]^T \in \mathbb{R}^n \tag{12}$$

where the logarithmic scale ψ is implemented in order to speed up the exploration of the design parameter searching. Each element of ψ is set to $\psi_i = 10^{p_i} (i = 1, 2 \ldots n)$ and the objective function become $J = [\, 10^{p_1} \quad 10^{p_2} \quad \ldots \quad 10^{p_n}\,]^T$, where n is a total number of control paraments.

Now, let consider the objective function $f(\boldsymbol{p}) = J(P, I, D, N, \alpha, \xi, \lambda)$ and the control parameters $p_i = \log \psi_i$. Next, determine the maximum number of iteration, $k_{\max}$. Then, employ the SED algorithm in (Sect. 3) to this optimization problem. After reach a maximum number of iteration $k_{\max}$, the optimal control parameters are obtained as $\boldsymbol{p}_{opt} = \bar{\boldsymbol{p}}(k_{\max})$. Finally, apply back the $\psi_{opt} = [\, 10^{p_{1opt}} \quad 10^{p_{2opt}} \quad \cdots \quad 10^{p_{n\,opt}}\,]$ into $C(s)$ and $NE(\boldsymbol{Er}(t), \Delta \boldsymbol{h}(t))$ controller in Fig. 1 to obtain the performance of control system.

5 TITO - Coupled Tank System

The liquid level of tank 1 $y_1(t) = h_1(t)$ and tank 2 $y_2(t) = h_2(t)$ are illustrated in Fig. 3. The plant modelling this TITO coupled tanks system is clearly described by [3]. The inputs bound of this system are $u_i(t) \in [0, 5](i = 1, 2)$.

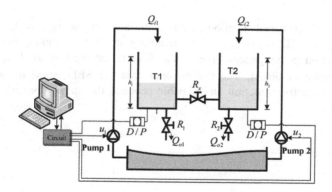

Fig. 3. Schematic of TITO coupled tank system.

The model system of the coupled tank with stochastic input delay as follows

$$\begin{bmatrix} y_1(s) \\ y_2(s) \end{bmatrix} = \begin{bmatrix} G_{11}(s)e^{-ds} & G_{12}(s)e^{-ds} \\ G_{21}(s)e^{-ds} & G_{22}(s)e^{-ds} \end{bmatrix} \begin{bmatrix} u_1(s) \\ u_2(s) \end{bmatrix} \tag{13}$$

where $d \in [1, 3]$ and

$$G_{11}(s) = \frac{0.08151s + 0.006356}{s^2 + 0.1402s + 0.003526}, \tag{14}$$

$$G_{12}(s) = \frac{2.637 \times 10^{-3}}{s^2 + 0.1402s + 0.003526}, \tag{15}$$

$$G_{21}(s) = \frac{2.966 \times 10^{-3}}{s^2 + 0.1402s + 0.003526}, \tag{16}$$

$$G_{22}(s) = \frac{0.07245s + 0.004507}{s^2 + 0.1402s + 0.003526}. \tag{17}$$

6 Numerical Example

The desired trajectories position of liquid levels is applied as shown in Table 1.

Table 1. Trajectories position of coupled tank

Tank	$t = 0$ s	$t = 500$ s	$t = 1000$ s	$t = 1500$ s
1	2	4	5	5
2	2	3	3	3

Here, the SED based method coefficients are set as $p_{max} = 4$, $p_{min} = -4$, $ET = 0.66, K_g = 0.022$, $w_1 = 4$, $w_2 = 5$, sampling time $t_s = 1$ s and $t_f = 1500$ s. The objective function convergence curve of the SED based method for $k_{max} = 2000$ in $p \in \mathbb{R}^{14}$ is shown in Fig. 4. The results show that the SED based method able to minimize the objective function and capable produce the stable convergence of the control system.

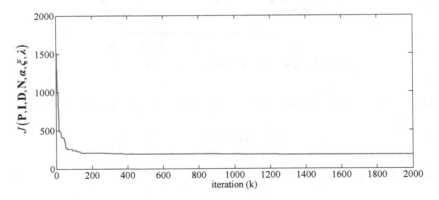

Fig. 4. The convergence curve of the SED based method.

Next, the initial neuroendocrine-PID control parameters $p(0)$ for $k = 0$ and the optimal control parameters p_{opt} for k_{max} are represented in Table 2. Note that, $p(0) \in \mathbb{R}^{14}$ is chosen based on several preliminary experiments. The statistical results of the SED based method after 30 trials are depicted in Table 3 in terms of the objective function, total norm error and the computational time. This finding shows that the SED based method successfully minimize the objective function and obtain the optimal neuroendocrine-PID controller parameters in order to control the liquid level tanks. Then, this fact also supports by the responses of liquid level for tank 1 and tank 2 that are shown in Figs. 5 and 6. The dot green line from both figures represents the reference of liquid level, the thick blue line represents the output liquid level tuning by SED while the thick red represents the initial parameter response before tuning. Notes that only the best output responses from 30 trials are plotted in the Figs. 5 and 6. Based on the results, it is clearly seen that the responses of data-driven neuroendocrine-PID based on SED produce almost zero steady-state error and lower overshoot. Moreover, this method also able to track the desired trajectory even though stochastic input delay occurred during the tuning process. Thus, it clarifies that this method is an effective method for the control of the coupled tank system with stochastic input delay. Then the SED algorithm also has good capability to tune the control parameters in presence of stochastic delay during the tuning process.

Table 2. Parameters design for SED based method

ψ	NE-PID Gain	$p(0)$	ψ corresponded to $p(0)$	p_{opt}	ψ corresponded to p_{opt}
ψ_1	P_{11}	0	1.000	0.9177	8.274
ψ_2	I_{11}	2.5	316.228	1.8275	67.220
ψ_3	D_{11}	0	1.000	-0.4783	0.332
ψ_4	N_{11}	0.4	2.512	-0.3281	0.470
ψ_5	α_{11}	0	1.000	0.57	3.715
ψ_6	ξ_{11}	0.8	6.310	1.3422	21.989
ψ_7	λ_{11}	0.4	2.512	0.8324	6.798
ψ_8	P_{22}	0.3	1.995	-0.4486	0.356
ψ_9	I_{22}	0.2	1.585	-0.1802	0.660
ψ_{10}	D_{22}	0.2	1.585	0.0976	1.252
ψ_{11}	N_{22}	0.5	3.162	0.448	2.805
ψ_{12}	α_{22}	0.5	3.162	0.2196	1.658
ψ_{13}	ξ_{22}	-0.2	0.631	-0.0825	0.827
ψ_{14}	λ_{22}	0.5	3.162	0.79	6.166

Table 3. Statistical result of TITO coupled tank with stochastic input delay

Algorithm		SED
Objective function $J(P, I, D, N, \alpha, \xi, \lambda)$	Mean	180.9328
	Best	175.6751
	Worst	202.3106
	Std.	5.9122
Norm-error $(\hat{e}_1 + \hat{e}_2)$	Mean	89.3133
	Best	40.509
	Worst	158.9257
	Std.	41.7591
Computational time (s)		766.6276

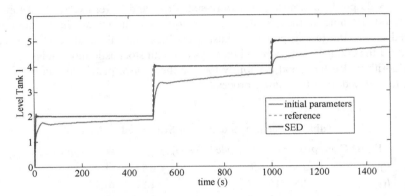

Fig. 5. Response of liquid level (tank 1) (Color figure online)

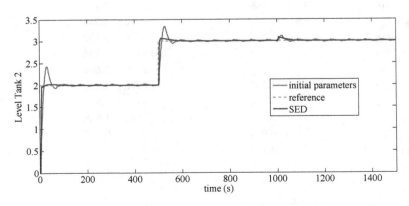

Fig. 6. Response of liquid level (tank 2) (Color figure online)

7 Conclusion

The performance of data-driven neuroendocrine-PID control based on the SED for TITO coupled tank system with stochastic input delay has been presented and discussed. Based on the statistical numerical example and outputs responses, the SED algorithm able to find the optimal PID control parameters and able to track the trajectories task given for stochastic input delay occurred during the tuning process. Therefore, it verifies that the fixed interval size characteristics in the SED based method that produce the independent gain sequence of iterations have enough energy to re-tune and finding the optimal neuro-endocrine-PID parameters when the delay occurs during the tuning process. In addition, this method also obtains good control accuracy without using any theoretical assumptions of plant modelling.

Acknowledgement. This study was partly supported by the Ministry of Higher Education, Government of Malaysia through the Fundamental Research Grant Scheme (FRGS) (RDU 160146) and Universiti Malaysia Pahang.

References

1. Ramli, M.S., Ahmad, M.A., Ismail, R.M.T.R.: Comparison of swarm adaptive neural network control of a coupled tank liquid level system. In: 2009 International Conference on Computer Technology and Development (ICCTD 2009), vol. 1, pp. 130–135 (2009)
2. Numsomran, A., Suksri, T., Thumma, M.: Design of 2-DOF PI controller with decoupling for coupled-tank process. In: International Conference on Control, Automation and System (ICCAS 2007), vol. 2, no. I, pp. 339–344 (2007)
3. Numsomran, A., Kangwanrat, S., Tipsuwannaporn, V.: Design of PI controller using MRAC techniques for coupled-tanks process. World Acad. Sci. Eng. Technol. **59**, 67–72 (2009)
4. Gupta, A., Goindi, S., Singh, G., Kumar, R.: Optimal design of PID controllers for time delay systems using genetic algorithm and simulated annealing. In: International Conference on Innovative Mechanism for Industry Applications (ICIMIA 2017), pp. 66–69 (2017)
5. Shu, H., Pi, Y.: PID neural networks for time-delay systems. Comput. Chem. Eng. **24**(2–7), 859–862 (2000)
6. Mingmei, W., Qiming, C., Yinman, C., Yingfei, W.: Model-free adaptive control method for nuclear steam generator water level. In: 2010 International Conference on Intelligent Computation Technology and Automation, pp. 696–699 (2010)
7. Xu, J., Hou, Z.: Notes on data-driven system approaches. Acta Autom. Sin. **35**(6), 668–675 (2009)
8. Hou, Z.S., Wang, Z.: From model-based control to data-driven control: survey, classification and perspective. Inf. Sci. (Ny) **235**, 3–35 (2013)
9. Ahmad, M.A., Azuma, S., Sugie, T.: Performance analysis of model-free PID tuning of MIMO systems based on simultaneous perturbation stochastic approximation. Expert Syst. Appl. **41**(14), 6361–6370 (2014)
10. Aksu, I. O., Coban, R.: Second order sliding mode control of MIMO nonlinear coupled tank system. In: 2018 14th International Conference on Advanced Trends Radioelecrtronics, Telecommunications and Computer Engineering, pp. 826–830 (2018)

11. Puralachetty, M.M., Pamula, V.K.: Differential evolution and particle swarm optimization algorithms with two stage initialization for PID controller tuning in coupled tank liquid level system. In: 2016 International Conference on Advanced Robotics and Mechatronics Differentiation, pp. 507–511 (2016)
12. Shukor, N.S.A., Ahmad, M.A.: Data-Driven PID Tuning Based on Safe Experimentation Dynamics for Control of Double-Pendulum-Type Overhead Crane. In: Hassan, M. (ed.) Intelligent Manufacturing & Mechatronics. LNME, pp. 295–308. Springer, Singapore (2018). https://doi.org/10.1007/978-981-10-8788-2_27
13. Ghazali, M.R., Ahmad, M.A., Falfazli, M., Jusof, M., Ismail, R.M.T. R.: A data-driven Neuroendocrine-PID controller for underactuated systems based on safe experimentation dynamics. In: 2018 IEEE 14th International Colloquium on Signal Processing and its Applications (CSPA 2018), March, pp. 9–10 (2018)
14. Ding, Y., Xu, N., Ren, L., Hao, K.: Data-driven neuroendocrine ultrashort feedback-based cooperative control system. IEEE Trans. Control Syst. Technol. 23(3), 1205–1212 (2015)
15. Marden, J.R., Ruben, S.D., Pao, L.Y.: A model-free approach to wind farm control using game theoretic methods. IEEE Trans. Control Syst. Technol. 21(4), 1207–1214 (2013)

Neuro-Fuzzy Sampling: Safe and Fast Multi-query Randomized Path Planning for Mobile Robots

Weria Khaksar[✉], Md Zia Uddin, and Jim Torresen

Robotics and Intelligent Systems Group, Department of Informatics,
University of Oslo, Blindern, 0316 Oslo, Norway
{weriak,mdzu,jimtoer}@ifi.uio.no

Abstract. Despite the proven advantages of probabilistic motion planning algorithms in solving complex navigation problems, their performance is restricted by the selected nodes in the configuration space. Furthermore, the choice of selecting the neighbor nodes and expanding the graph structure is limited by a set of deterministic measures such as Euclidean distance. In this paper, an improved version of multi-query planners is proposed which utilizes a neuro-fuzzy structure in the sampling stage to achieve a higher level of effectiveness including safety and applicability. This planner employs a set of expert rules concerning the distance to the surrounding obstacles and constructs a fuzzy controller. Then, parameters of the resulting fuzzy system are optimized based on a hybrid learning technique. The outcome of the neuro-fuzzy system is implemented on a multi-query planner to improve the quality of the selected nodes. The planner is further improved by adding a post-processing step which shortens the path by removing the redundant segments and by smoothing the resulting path through the inscribed circle of any two consecutive segments. The planner was tested through simulation in different planning problems and was compared to a set of benchmark algorithms. Furthermore, the proposed planner was implemented on a robotic system. Simulation and experimental results indicate the superior performance of the planner in terms of safety and applicability.

Keywords: Path planning · Mobile robot · Sampling-based · Multi-query · Safety · Neuro-fuzzy system

1 Introduction

In the field of motion planning, sampling-based planners have been successfully applied to solve difficult problems in high-dimensional spaces. These algorithms are unique in the fact that planning occurs by sampling the configuration space. Original sampling-based planners such as Probabilistic Roadmaps (PRM) [1, 2], Rapidly-Exploring Random Trees (RRT) [3, 4], and Expansive Space Trees (EST) [5], are proven to be probabilistically complete as the probability of finding a solution in these planners is one when the input size goes to infinity. These algorithms have been improved further to achieve some form of optimality in the generated solutions.

© Springer Nature Singapore Pte Ltd. 2019
J.-H. Kim et al. (Eds.): RiTA 2018, CCIS 1015, pp. 13–27, 2019.
https://doi.org/10.1007/978-981-13-7780-8_2

Optimal sampling-based planners such as PRM* and RRT* [6, 7] are asymptotically optimal as the solutions found by these algorithms converge asymptotically to the optimum, if one exists, with the probability one as the input size goes to infinity.

Besides the initial objective of these algorithms which was to reduce the planning cost and to quickly produce a collision-free path, recent application domains require different planning objectives such as optimality, and safety. Furthermore, these methods should be applicable to different robotic systems with different physical and kinematic constraints. In some of the recent applications of robotics such as health care [8], rescue [9], and logistics [10], the motion planning algorithm needs to provide safe navigation paths where the robot is moving as far away as possible from the obstacles, and also the path planner should be able to quickly provide solutions for the query. Furthermore, the kinematic constraints of the robot should be considered within the path planning module. Especially when dealing with nonholonomic robots, the curvature of the generated paths plays an important role in the successful implementation of the result.

In this paper, an extension of the Probabilistic Roadmaps (PRM) is proposed which is able to detect safe navigation domains in the planning space and quickly generate navigation plans within those regions. Additionally, a simple and effective post-processing technique is introduced which removes the redundant segments of the resulting path and improves the curvature of the generated path to match the physical constraints of the robot. Figure 1 shows the results obtained by the proposed method versus the original PRM planner in a 2D planning space.

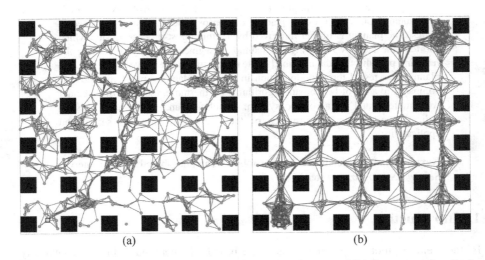

(a)   (b)

Fig. 1. Different performances of (a) the original PRM planner and (b) the proposed method. Both algorithms result in a path (red curves) from the start position (yellow square) and the goal position (green square). Obstacles are illustrated by the black blocks. (Color figure online)

First, a neuro-fuzzy sampling controller is designed which receives the positional information about each generated sample and decides whether to accept the sample or not. This sampling controller aims in biasing the samples towards area close to the initial and final positions and far away from the obstacles. A set of 10^3 different planning environments were designed and 10^2 samples were generated in each environment to provide enough training data for the proposed neuro-fuzzy structure which consists of a neural network with six layers for constructing and optimizing the fuzzy controller, and a fuzzy sampling controller to decide about accepting or rejecting each generated sample. The resulting set of samples is then used to learn the planning space through the probabilistic roadmaps structure. Another criterion is also considered in the sampling process which defines a forbidden sampling area around each sample and no further sampling takes place within that region to ensure a diverse and sparse set of samples. This method follows the low-dispersion sampling strategy. Next, the post-processing component of the proposed method is applied on the resulting path which first, removes the redundant segments of the initial solution and then, smooths the path based on the inscribed circle of any two consecutive segments of the path.

The rest of the paper is organized as follows: Sect. 2 provides basic information about the PRM planner while Sect. 3 describes the low-dispersion approach. The structure of the proposed neuro-fuzzy system is discussed in Sect. 4 while the introduced post-processing methods are explained in Sect. 5. Section 6 includes the simulation and experimental results and finally the work in summarized in Sect. 7.

2 The PRM Planner

Most of the sampling-based navigation algorithms perform the path planning task in the configuration space which is the space of all possible transformations that could be applied to a robot. The concept of configuration space has been introduced [11] to simplify complex planning situations in the actual workspace of a robot. The main idea behind the PRM planner is to learn the obstacle-free space and the connectivity of the configuration space by randomly selecting several configurations and then connecting them according to a given strategy to construct a graph. The resulting graph can be used for solving any given feasible queries [12]. The main advantage of this planner is the ability to learn the space without requiring a detailed map which is sometimes computationally impossible to get. Instead, it relies on two main components. The first component is responsible to determine if a given configuration is in collision with an obstacle or not, while the second component, i.e. the local planner, determines if the robot can freely move from one configuration to another without any collision. There are numerous extensions of the PRM algorithm focusing on different objectives such as optimality, rapidity, effectiveness, and efficiency. Lazy PRM planner was proposed to increase the planning speed by delaying the collision-check as much as possible [13]. Low dispersion version of PRM (LD-PRM) was introduced to speed up the planning process by reducing the density of the graph [14]. Randomized bridge builder (RBB) [15] and the Gaussian PRM [16] were proposed to increase the efficiency of the PRM in narrow spaces. Recently, the asymptotic optimal form of PRM, i.e. PRM* has been introduced which improves the quality of the generated solutions [6].

3 Low-Dispersion Sampling

In this work, a randomized multi-query path planning algorithm is proposed for safe and quick navigation based on the LD-PRM algorithm [14] and first, this algorithm needs to be briefly introduced. The main motivation of the low dispersion-PRM (LD-PRM) planner is to improve the dispersion of the generated samples by the original PRM algorithm so that the planner can capture the connectivity of the workspace and solve planning queries with fewer samples, which reduces the processing time of the algorithm. This algorithm works similarly to the original PRM with one major difference in the learning phase. In LD-PRM, the learning phase is different such that the generated samples need to fulfill an additional criterion in order to be included in the roadmap. This criterion implies that the samples are forbidden to be close to each other more than a predefined radius. Based on the Lebesgue measure (volume) of the collision-free configuration space, this radius can be defined as follows:

$$R_S(n) = \left[L\left(Q_{free}\right)(n - \lambda)/\left(\pi n^2\right)\right]^{1/2} \tag{1}$$

Where Q_{free} is the free configuration space, n is the number of samples (nodes) in the roadmap, and λ is a positive scaling scalar. The idea behind this formulation is that if we want to fit non-overlapping balls inside the configuration space, the aggregate volume of these balls should be smaller than the volume of the free space. More information about the LD-PRM planner can be found in [14]. This algorithm can be summarized as follows:

THE LOW-DISPERSION PRM PLANNER.

Inputs:	N (number of nodes in the roadmap)
	k (number of closest neighbors for each node)
	R (radius of the forbidden regions)
Output:	A dispersed roadmap $RM = (V, E)$

```
1:   V ← ∅,    E ← ∅;
2:   While |V| < N do
3:      Repeat
4:         x ← a random configuration
5:      Until x is collision-free
6:      If for all x' ∈ V:  d(x, x') ≥ R
7:              V ← V ∪ {x}
8:      End if
9:   End While
10:  For all x ∈ V, do
11:     Nx ← the k closest neighbors of x from V
12:     For all x' ∈ Nx do
13:        If (x, x') ∉ E and (xx') is collision-free then
14:              E ← E ∪ {(x, x')}
15:        End if
16:     End for
17:  End for
```

4   Neuro-Fuzzy Sampling Controller

In order to make safe navigation plans, the LD-PRM planner needs to detect safe navigation passages in the planning space and also, quickly planning the path to avoid generating high-cost solutions. To achieve these objectives, a fuzzy logic sampling controller is proposed which receives each sample's coordinate (x_s, y_s), and calculated three planning variables including the distance from the sample to the closest obstacle (D_{obs}), and the distance of the sample to the start (D_s) and the goal (D_g) positions of the planning query. These variables are required to ensure safe and fast path planning. Accordingly, the proposed fuzzy logic system calculates the output which follows a step function and equals to zero if the sample does not meet the planning criteria and equals to one otherwise. If the output of a sample is zero, the sample is rejected and will be discarded from the graph and if the output is one, the sample is accepted and added to the graph. The purpose of this fuzzy logic system is to find samples which are far from the obstacles. Furthermore, the planner aims to generate more samples in the vicinity of the start and goal positions to ensure maximum success in the path planning task. Figure 2 shows the effect of the fuzzy logic sampling controller on the generated samples.

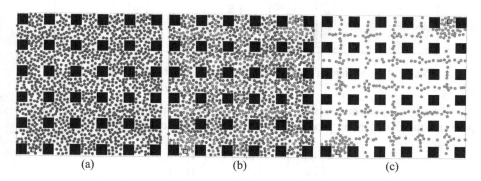

Fig. 2. The effect of the proposed fuzzy structure on the distribution of the samples. (a) The initial samples generated by the LD-PRM planner, (b) the fuzzy logic sampling controller accepts suitable samples (green dots) and rejects unfit samples (red dots), and (c) the final distribution of the accepted samples. The density of the samples is higher around start and goal positions shown by yellow and green squares respectively. (Color figure online)

The next step is to design the fuzzy logic sampling controller. An adaptive neuro-fuzzy inference system (ANFIS) [17, 18] is utilized which can construct and optimize the fuzzy logic structure. The construction of the fuzzy logic system is done using the subtractive clustering method [19] and then the parameters of the constructed fuzzy logic system will be optimized through the neural network structure by means of a hybrid technique of back-propagation algorithm and the least square adaptation method [20]. The proposed ANFIS structure follows a systematic approach for generating fuzzy rules and utilizes a first-order Sugeno fuzzy system with the following rule structure:

$$\text{IF } D_s \text{ is } A_1 \text{ AND } D_g \text{ is } A_2 \text{ AND } D_{obs} \text{ is } A_3 \tag{2}$$

$$\text{THEN } \left\langle \begin{array}{l} v = f_1\left(D_s, D_g, D_{obs}, \theta_s, \theta_g, \theta_{obs}\right) \\ \omega = f_2\left(D_s, D_g, D_{obs}, \theta_s, \theta_g, \theta_{obs}\right) \end{array} \right. \tag{3}$$

where, f_1 and f_2 are first-order polynomial functions of the input variables. The proposed ANFIS is represented by a six-layer feedforward neural network as illustrated in Fig. 3.

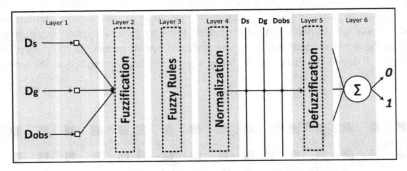

Fig. 3. The structure of the proposed adaptive neuro-fuzzy inference system.

In the first layer, neurons simply pass external crisp signals to layer 2 while layer 2 performs the fuzzification procedure with a bell-shaped activation function.

$$\begin{bmatrix} out_1^1 \\ out_2^1 \\ out_3^1 \end{bmatrix} = \begin{bmatrix} in_1^1 \\ in_2^1 \\ in_3^1 \end{bmatrix} = \begin{bmatrix} D_s \\ D_g \\ D_{obs} \end{bmatrix}, \tag{4}$$

$$out_i^2 = 1 / \left[1 + \left(\frac{in_i^2 - a_i}{c_i}\right)^{2b_i}\right], \tag{5}$$

where a_i, b_i, and c_i are control parameters for the center, width, and slope of the activation function for neuron i, respectively. Each neuron in the third layer represents a single fuzzy rule which receives the fuzzified output from the previous neuron and calculates the firing strength of the corresponding rule. As suggested in the original ANFIS structure [17, 20], the conjunction of the rule antecedents is assessed by the product operator.

$$out_i^3 = \prod_{j=1}^{k} in_{ji}^3,$$

(6)

The resulting values are normalized in the next layer by computing the normalized firing strength of each rule as the ratio of the firing strength of each rule to the sum of the firing strength for all rules.

$$out_i^4 = in_{ii}^4 / \sum_{j=1}^{n} in_{ji}^4 = \mu_i / \sum_{j=1}^{n} \mu_j = \overline{\mu_i},$$

(7)

Where the value of μ_j is the firing strength of neuron i. Layer 5 defuzzifies the resulting values from the previous layer. Each defuzzification neuron computes the weighted consequent value of a given rule as:

$$out_i^5 = \overline{\mu_i}[k_{i0} + k_{i1}x1 + k_{i2}x2],$$

(8)

Here, k_{i0}, k_{i1}, and k_{i2} are the parameters of the ith rule. At the last layer, a single summation neuron computes the sum of outputs of all previous defuzzification neurons and produces the final output of the network.

$$out^6 = \sum_{i=1}^{n} in_i^6 = \sum_{i=1}^{n} \overline{\mu_i}[k_{i0} + k_{i1}x1 + k_{i2}x2],$$

(9)

Now the ANFIS is ready for training and in order to provide sufficient training data for the neuro-fuzzy system, a set of 10^3 different planning environments were designed and 10^2 samples were created in each one. Then, each sample was given an associated output having a Boolean value (0 or 1), based on four simple rules as given in Table 1. Three parameters were used to generate the dataset including λ_s, λ_g, and λ_{obs} which represent the minimum distance from the start position, minimum distance to the goal position, and minimum safety distance from obstacles. These parameters were used to assign proper output value to each sample.

Table 1. Four simple rules used for generating the training dataset.

D_s	D_g	D_{obs}	Sample output
$< \lambda_s$	$> \lambda_g$	–	1
$> \lambda_s$	$< \lambda_g$	–	1
–	–	$> \lambda_{obs}$	1
Otherwise			0

Figure 4 shows 10^2 samples generated in a simple planning space. Different values of the generated data and the corresponding deviations are illustrated for both input and output values. As mentioned before, this process was repeated in 10^3 different planning spaces with different arrangements of the obstacles. These environments were created entirely based on random numbers to generalize the behavior of the fuzzy logic sampling controller. The corresponding output values are presented in Fig. 5.

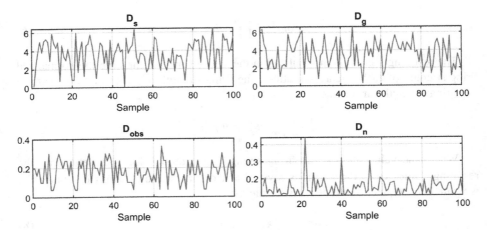

Fig. 4. An instance of the input training dataset generated in an environment with a random arrangement of obstacles and for a random planning query.

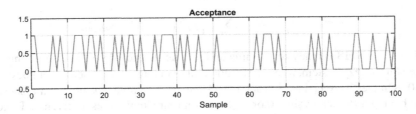

Fig. 5. Corresponding output values in one environment according to the four given rules.

After generating 10^5 samples (10^2 samples in 10^3 spaces), the initial fuzzy logic structure was created with the subtractive clustering method. Then, the optimization process took place using the hybrid technique. The optimized membership functions are shown in Fig. 6 and the corresponding training curve is presented in Fig. 7.

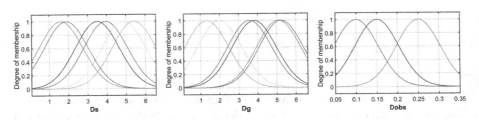

Fig. 6. The input membership functions after the construction and optimization by the proposed neuro-fuzzy structure.

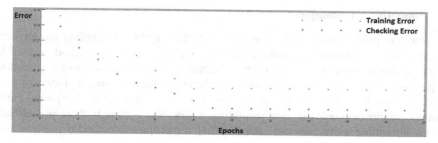

Fig. 7. The training and checking errors in 20 epochs. The training error reaches down to 0.1393 while the checking error reaches the minimum value of 0.1257.

As can be observed from the training and checking curves, the training error reaches its minimum after eight epochs while the checking error keeps decreasing until epoch nine. The obtained values which were used in the final planner were based on the training for eight epochs to avoid overfitting in the training process.

5   Post-processing Approach

To improve the quality of the generated path by the proposed planner, two post-processing techniques are used to shorten the length of the path and to smooth the path for real implementations as presented in the following sections.

5.1   Path Shortening

Given the path resulting from the planner, the path shortening process removes the redundant segments of the path by checking the availability of a direct connection between any two nodes on the path, starting from the initial node on the path. As shown in Fig. 8(b), after removing the redundant segments, no direct connection can be made between any pair of non-adjacent nodes on the path.

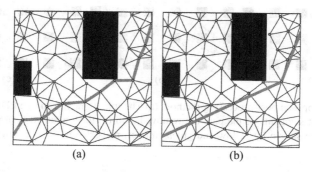

(a)   (b)

Fig. 8. An instance of the path reduction process. (a) A segment of the initial path and (b) the reduced path after the first post-processing process.

5.2 Smoothing Technique

The resulting path from the path shortening is still incapable of handling the physical limitation of a nonholonomic robotic system. A path smoothing method is used to overcome this issue by finding the inscribed circle of any two consecutive segments of the path as shown in Fig. 9. After computing the center location and the radius of each circle, the middle node will be removed, and the path is updated by the partial curve from the circle between the two remaining nodes. Even though there are already various smoothing techniques available [21–24], the proposed method is implemented since it requires lower processing resources.

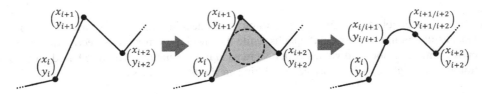

Fig. 9. The performance of the proposed path smoothing method. By finding the inscribed circle of any given two segments, the middle curve will be added to the path and replace the sharp edges between the two segments.

Figure 10 shows the summary of the post-processing module where the initial, shortened and smoothed paths are shown in a simple 2D environment with 36 convex obstacles.

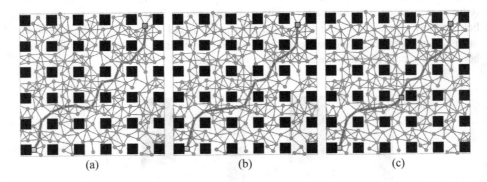

Fig. 10. The proposed post-processing method. (a) The initial graph from the LD-PRM planner, (b) the redundant segments of the path are removed, and (c) the smoothed final path. Start and final positions are shown by yellow and green squares, respectively. (Color figure online)

6 Results and Discussion

To evaluate the performance of the proposed path planning approach and to compare it with the original PRM-based planners, several simulation and experimental studies have been conducted as presented in the following sections.

6.1 Simulation Studies

The proposed approach was simulated in three different planning spaces with different placements of the obstacles as illustrated in Fig. 11. The simulation was programmed in MatLab R2017a on a desktop with a 3.40-GHz Intel Core i7 processor and 32 GB of memory. To have a general and reliable overview of the performances, each planner was tested for 100 iterations per problem.

Fig. 11. Simulation of the proposed method in three test planning problems with different arrangements of the obstacles.

The planner was able to successfully generate safe navigation paths while maintaining the running time and the solution cost as compared to other two counterparts including the original PRM and the LD-PRM planners. The numerical comparative results are shown in Table 2 and Fig. 12 in terms of path length (PL), processing time (RT), and the minimum distance to the obstacles (D_{min}). The proposed planner provides safe navigation plans while avoiding any increase in the solution cost and processing time.

Table 2. Comparative simulation results. The results are averaged over all three environments and 100 iterations per planner per environment. Standard deviations (Std.) are also included to see the stability of the results.

Test	PRM			LD-PRM			Proposed method		
	PL (m)	RT (sec)	D_{min} (m)	PL (m)	RT (sec)	D_{min} (m)	PL (m)	RT (sec)	D_{min} (m)
Mean	88.82	34.53	0.26	85.00	28.32	0.24	85.15	28.70	1.10
Std.	4.35	4.40	0.08	2.66	4.19	0.08	2.81	4.97	0.18

To conclude the simulation results, the proposed planner keeps the low processing time property of the LD-PRM and therefore performs better than the original PRM. It also provides safer navigation plans while maintaining the solution cost (path length).

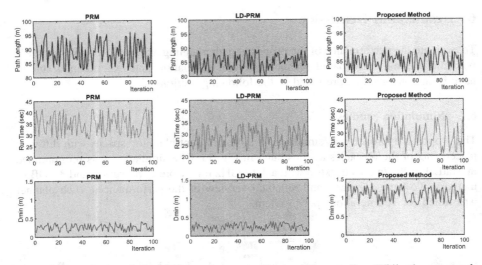

Fig. 12. Numerical analysis of the performances in simulation studies. While the proposed planner maintains the same level of path lengths with the LD-PRM, and the same level of processing time with PRM and LD-PRM, it significantly improves the safety.

6.2   Experimental Setup

The planner was implemented on a nonholonomic differential drive robotic platform to test the actual performances. The main control platform for the robot was robotic operating system (ROS). The details of the robotic platform are summarized in Table 3. Two test problems are designed with two different arrangements of the obstacles as can be seen in Fig. 13. The first case is a scattered arrangement of twelve obstacles in a 3 m × 4 m plane while the second arrangement is a maze in the same plane with six obstacles. The numerical results of these two experiments are summarized in Table 4.

Table 3. The experimental setup configuration.

Component	Specifications
Robot	TurtleBot3 (Burger), 138 mm × 178 mm × 192 mm $v_{max} = 0.22$ (m/s), $v_{max} = 2.84$ (rad/s)
Actuator	Dynamixel XL430-W250
Sensor	360 Laser Distance Sensor LDS-01
SBC	Raspberry Pi 3 Model B
MCU	32-bit ARM Cortex®-M7 with FPU (216 MHz, 462 DMIPS)
Power source	Lithium polymer 11.1 V 1800mAh/19.98Wh 5C

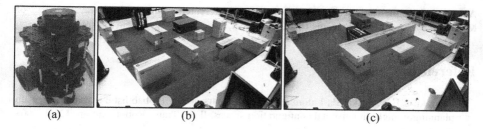

(a)   (b)   (c)

Fig. 13. The experimental configuration. (a) The robotic platform which is a turtlebot3, burger model, (b) and (c) two test setups with different placements of the obstacles.

Table 4. Numerical results obtained from the experiments.

Environment	PL (m)	RT (sec)	D_{min} (m)	λ_c (1/sec)
Scattered	8.27	4.52	0.21	0.83
Maze	9.16	6.09	0.17	0.56

The presented numerical results are shown in terms of path length (PL) as the total traveled distance by the robot, processing time (RT), minimum distance to the obstacles (D_{min}), and the overall control frequency (λ_c) as the number of control commands given to the robot per second. The planner managed to keep the processing time as low as 6 (sec) while providing safe trajectories where the robot was at least 17(cm) away from the obstacles.

7 Summary

In this paper, a multi-query path planning algorithm was proposed for planning safe navigation for mobile robotic systems based on the low-dispersion probabilistic roadmaps (LD-PRM) method. The core of the proposed method is an adaptive neuro-fuzzy inference system (ANFIS) which constructs and optimizes the parameters of the fuzzy sampling controller. Then the fuzzy system was implemented on the sampling module to select the samples according to the safety criterion. The resulted path was further improved by a post-processing step which removes the redundant segments of the path as well as the sharp edges. The resulting reduced and smoothed paths can guide a mobile robot among different arrangements of the obstacles within the planning space. The proposed method was tested through several simulation, comparative, and experimental studies. The superiority of the approach in generating safe navigation solutions was concluded while the solution cost and processing time were maintained without any meaningful increase.

This work can further be improved by using other optimization techniques for designing the fuzzy sampling controller. In addition, other types of planning such as dynamic motion planning or maples navigation can be considered for further advancements of the proposed approach.

Acknowledgment. This work is supported by the Research Council of Norway as a part of Multimodal Elderly Care Systems (MECS) project, under grant agreement 247697.

References

1. Kavraki, L.E., Svestka, P., Latombe, J.C., Overmars, M.H.: Probabilistic roadmaps for path planning in high-dimensional configuration spaces. IEEE Trans. Robot. Autom. **12**(4), 566–580 (1996)
2. Kala, R.: Increased visibility sampling for probabilistic roadmaps. In: 2018 IEEE International Conference on Simulation, Modeling, and Programming for Autonomous Robots (SIMPAR), pp. 87–92 (2018)
3. LaValle, S.M., Kuffner, J.J.: Randomized kinodynamic planning. Int. J. Robot. Res. **20**(5), 378–400 (2001)
4. Neto, A.A., Macharet, D.G., Campos, M.F.M.: Multi-agent rapidly-exploring pseudo-random tree. J. Intell. Robot. Syst. **89**(1–2), 69–85 (2018)
5. Hsu, D., Latombe, J., Motwani, R.: Path planning in expansive configuration spaces. In: Proceedings of International Conference on Robotics and Automation, vol. 3, pp. 2719–2726 (1997)
6. Karaman, S., Frazzoli, E.: Sampling-based algorithms for optimal motion planning. Int. J. Robot. Res. **30**(7), 846–894 (2011)
7. Schmerling, E., Janson, L., Pavone, M.: Optimal sampling-based motion planning under differential constraints: the driftless case. In: 2015 IEEE International Conference on Robotics and Automation (ICRA), pp. 2368–2375 (2015)
8. Bemelmans, R., Gelderblom, G.J., Jonker, P., de Witte, L.: Socially assistive robots in elderly care: a systematic review into effects and effectiveness. J. Am. Med. Dir. Assoc. **13** (2), 114–120.e1 (2012)
9. Yazdani, F., Brieber, B., Beetz, M.: Cognition-enabled robot control for mixed human-robot rescue teams. In: Menegatti, E., Michael, N., Berns, K., Yamaguchi, H. (eds.) Intelligent Autonomous Systems 13, vol. 302, pp. 1357–1369. Springer, Cham (2016). https://doi.org/10.1007/978-3-319-08338-4_98
10. Perdoch, M., Bradley, D.M., Chang, J.K.: Herman, H., Rander, P., Stentz, A.: Leader tracking for a walking logistics robot. In: 2015 IEEE/RSJ International Conference on Intelligent Robots and Systems (IROS), pp. 2994–3001 (2015)
11. LaValle, S.M.: Planning Algorithms. Cambridge University Press, Cambridge (2006)
12. Tsianos, K.I., Sucan, I.A., Kavraki, L.E.: Sampling-based robot motion planning: towards realistic applications. Comput. Sci. Rev. **1**(1), 2–11 (2007)
13. Bohlin, R., Kavraki, L.E.: Path planning using lazy PRM. In: Proceedings 2000 ICRA. Millennium Conference. IEEE International Conference on Robotics and Automation. Symposia Proceedings (Cat. No.00CH37065), vol. 1, pp. 521–528 (2000)
14. Khaksar, W., Hong, T.S., Khaksar, M., Motlagh, O.: A low dispersion probabilistic roadmaps (LD-PRM) algorithm for fast and efficient sampling-based motion planning. Int. J. Adv. Robot. Syst. **10**(11), 397 (2013)
15. Hsu, D., Jiang, T., Reif, J., Sun, Z.: The bridge test for sampling narrow passages with probabilistic roadmap planners. In: 2003 IEEE International Conference on Robotics and Automation (Cat. No. 03CH37422), vol. 3, pp. 4420–4426 (2003)
16. Boor, V., Overmars, M.H., van der Stappen, A.F.: The Gaussian sampling strategy for probabilistic roadmap planners. In: Proceedings 1999 IEEE International Conference on Robotics and Automation (Cat. No. 99CH36288C), vol. 2, pp. 1018–1023 (1999)

17. Jang, J.S.R.: ANFIS: adaptive-network-based fuzzy inference system. IEEE Trans. Syst. Man Cybern. **23**(3), 665–685 (1993)
18. Chen, Y., Chang, C.: An intelligent ANFIS controller design for a mobile robot. In: 2018 IEEE International Conference on Applied System Invention (ICASI), pp. 445–448 (2018)
19. Chiu, S.L.: Fuzzy model identification based on cluster estimation. J. Intell. Fuzzy Syst. **2**(3), 267–278 (1994)
20. Karaboga, D., Kaya, E.: Adaptive network based fuzzy inference system (ANFIS) training approaches: a comprehensive survey. Artif. Intell. Rev. 1–31 (2018)
21. Fleury, S., Soueres, P., Laumond, J.P., Chatila, R.: Primitives for smoothing mobile robot trajectories. IEEE Trans. Robot. Autom. **11**(3), 441–448 (1995)
22. Ravankar, A., Ravankar, A.A., Kobayashi, Y., Emaru, T.: Path smoothing extension for various robot path planners. In: 2016 16th International Conference on Control, Automation and Systems (ICCAS), pp. 263–268 (2016)
23. Su, K.-H., Phan, T.-P.: Robot path planning and smoothing based on fuzzy inference. In: 2014 IEEE International Conference on System Science and Engineering (ICSSE), pp. 64–68 (2014)
24. Yang, K., Sukkarieh, S.: An analytical continuous-curvature path-smoothing algorithm. IEEE Trans. Robot. **26**(3), 561–568 (2010)

Hierarchical Attention Networks for Different Types of Documents with Smaller Size of Datasets

Hon-Sang Cheong[1(✉)], Wun-She Yap[1], Yee-Kai Tee[1], and Wai-Kong Lee[2]

[1] Lee Kong Chian Faculty of Engineering and Science,
Universiti Tunku Abdul Rahman, Sungai Long, Malaysia
dexter6855@hotmail.com, {yapws, teeyk}@utar.edu.my
[2] Faculty of Information and Communication Technology,
Universiti Tunku Abdul Rahman, Sungai Long, Malaysia
wklee@utar.edu.my

Abstract. The goal of document classification is to automatically assign one or more categories to a document by understanding the content of a document. Much research has been devoted to improve the accuracy of document classification over different types of documents, e.g., review, question, article and snippet. Recently, a method to model each document as a multivariate Gaussian distribution based on the distributed representations of its words has been proposed. The similarity between two documents is then measured based on the similarity of their distributions without taking into consideration its contextual information. In this work, a hierarchical attention network (HAN) which can classify a document using the contextual information by aggregating important words into sentence vectors and the important sentence vectors into document vectors for the classification was tested on four publicly available datasets (TREC, Reuter, Snippet and Amazon). The results showed that HAN which can pick up important words and sentences in the contextual information outperformed the Gaussian based approach in classifying the four public datasets consisting of questions, articles, reviews and snippets.

Keywords: Document classification · Machine learning · Hierarchical attention network · Accuracy · Dataset

1 Introduction

Document classification is one of the research area in natural language processing. The goal of document classification is to automatically assign one or more categories to a document by understanding the content of a document. Due to the massive usage of cloud storages, data analytics tools have been incorporated by data storage vendors into

Supported by the Collaborative Agreement with NextLabs (Malaysia) Sdn Bhd (Project title: Advanced and Context-Aware Text/Media Analytics for Data Classification).

© Springer Nature Singapore Pte Ltd. 2019
J.-H. Kim et al. (Eds.): RiTA 2018, CCIS 1015, pp. 28–41, 2019.
https://doi.org/10.1007/978-981-13-7780-8_3

their products. The availability of the large amount of data have motivated the development of efficient and effective document classification techniques. Such document classification techniques found their applications in topic labelling [24], sentiment classification [16], short-text categorization [7] and spam detection [11].

Words in a document can be represented as embeddings that act as feature for document classification. A document with smaller size, also known as short text may not be recognized easily as compared to longer text due to the issue of data sparsity and ambiguity. Traditional approach to classify a short text is to represent texts with bag-of-word (BoW) vectors. Even though BoW is simple and straightforward, but it did not take consideration on the contextual information of the document.

To utilize more contextual information in a document or text, biterm topic model (BTM) [4] had been introduced. However, BTM model may suffer from curse of dimensionality problem due to the use of sparse vector representation. In recent years, much research has been devoted to tackle the curse of dimensionality by learning distributed representation of words in documents. A neural probabilistic language model was proposed by Bengio et al. [2] to learn a distributed representation for words that capture neighboring sentences semantically. Instead of using a neural probabilistic language model, Mnih and Kavukcuoglu [17] used training log-bilinear models with noise-contrastive estimation too learn word embeddings. They also found that the embeddings learned by the simpler models can perform at least as well as those learned by the more complex one. Along this direction, Pennington, Socher and Manning [19] proposed the use of global log-bilinear regression model that efficiently leverages statistical information by training only on the nonzero elements in a word-word co-occurrence matrix.

On the other hand, deep learning approaches had been proposed for document classification. These approaches include convolutional neural network (CNN) [10] and recurrent neural network (RNN) based on long short term memory (LSTM) [8]. Even though neural network based text classification approaches had been proved to be quite effective by Kim [10] and Tang et al. [23] independently, the annotation of each word in these approaches only summarizes the preceding words, but never consider the following words. Hence, Bahdanau et al. [1] proposed the use of a bi-directional RNN that considers both the preceding and following words.

Recently, Yang et al. [25] proposed a new approach based on deep learning. They named the new approach as hierarchical attention network (HAN). The intuition of the proposed model is simple where a document can be split into sentences and each sentence can be split into words. Thus, the HAN model was designed to capture these two levels that form a document. Instead of using bi-directional RNN, HAN used bi-directional gated recurrent unit (GRU). GRU is a new type of hidden unit proposed by Cho et al. [5]. GRU is inspired by the LSTM unit with a simpler structure that can be implemented easily. First, the preceding and following words in each sentence are considered by HAN model such that the more important words will be given higher weightage. Subsequently, the preceding and following sentences in a document are considered such that the more important sentence will be given higher weightage. Finally, a document is classified based on such weightages. The efficiency of HAN had been proved against six publicly available datasets (Yelp 2013 [23], Yelp 2014 [23], Yelp 2015 [23], Yahoo Answer [26], IMDB Review [6] and Amazon [26]). Besides, Poon et al. [21] also demonstrated that HAN model is suitable for document level polarity classification.

All six datasets tested with HAN share the same similarities: (1) All documents are of the same type – user review; (2) Total number of documents in each dataset is large ranging from 335,018 to 3,650,000; (3) The size of vocabulary in each dataset is large ranging from 211,245 to 1,919,336. As short text classification is different with normal text classification due to the issue of data sparsity and ambiguity [7], thus the applicability of HAN on short text classification remains unclear.

Recently, Gaussian model proposed by Rousseau et al. [22] had demonstrated its effectiveness in recognizing short texts. The Gaussian method models each document as a multivariate Gaussian distribution based on the distributed representations of its words. The similarity between two documents is then measured based on the similarity of their distributions without taking into consideration its contextual information. However, contextual information is not taken into consideration in the proposed Gaussian model. To exploit the contextual information, a short text categorization strategy based on abundant representation was proposed by Gu et al. which subsequently outperformed the Gaussian model over a public dataset, Snippet [20].

In this paper, our contributions are listed as follows:

1. The efficiency of HAN model against different types of documents with smaller datasets and vocabulary is investigated.
2. The accuracy results of HAN against four selected datasets that consist of different types of documents are slightly better than state-of-the-art document classification methods.

2   The Hierarchical Attention Network Proposed by Yang et al.

A hierarchical attention network (HAN) was proposed by Yang et al. [25] for document classification with two unique characteristics: (1) It has a hierarchical structure that mirrors the hierarchical structure of documents; (2) it has two levels of attention mechanisms that applied at the word- and sentence-level to capture qualitative information when classifying a document. Figure 1 shows the architecture of HAN.

Assume that a document has n sentences s_i and each sentence contains m words. Let w_{it} with $t = 1, \ldots, m$ denotes the t-th word in the i-th sentence. HAN consists of the following components:

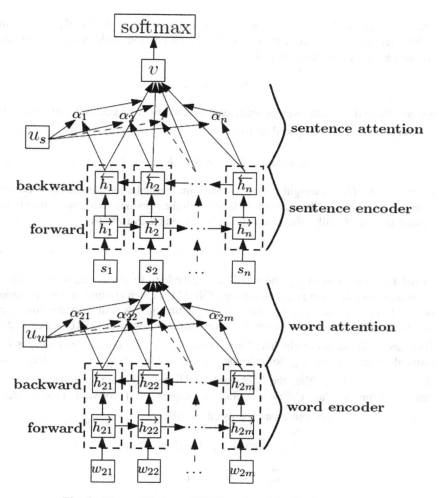

Fig. 1. The architecture of HAN proposed by Yang et al. [25]

1. **Gated Recurrent Unit (GRU) Based Sequence Encoder:** Cho et al. [5] proposed a new type of hidden unit, namely GRU that is inspired by the LSTM unit [8]. Differ with LSTM that has a memory cell and four gating units, GRU consists of two gating units only leading to simpler implementation and computation. The two gating units are named as reset gate r_t and update gate z_t for t-th hidden unit. These two gating units adaptively control the information ow inside the unit. First, the reset gate r_t is computed as follows:

$$r_t = \sigma(W_r x_t + U_r h_{t-1} + b_r) \tag{1}$$

where σ is the logistic sigmoid function, x_t is the input at time t, W_r and U_r are weight matrices which are learned and h_{t-1} is the previous hidden state. Then, the update gate z_t is computed as follows:

$$z_t = \sigma(W_z x_t + U_z h_{t-1} + b_z) \tag{2}$$

where W_z and U_z are weight matrices which are learned. Subsequently, the candidate state $\widetilde{h}_t$ is computed as follows:

$$\widetilde{h}_t = tanh(W_h x_t + r_t \odot U_h h_{t-1} + b_h) \tag{3}$$

where W_h and U_h are weight matrices which are learned and $\odot$ is the element-wise multiplication. Notice that r_t controls how much the previous state contributes to the candidate state. Finally, the new state h_t is computed as follows:

$$h_t = (1 - z_t) \odot h_{t-1} + z_t \odot \widetilde{h}_t \tag{4}$$

2. **Word Encoder:** Given w_{it}, the words are embedded to vectors $x_{it} = W_e w_{it}$ where W_e is an embedding matrix. Bidirectional GRU [1] is then applied to get annotation of words by summarizing not only the preceding words, but also the following words. A bidirectional GRU consists of forward and backward GRU's, denoted as $\overrightarrow{GRU}$ and $\overleftarrow{GRU}$ respectively. The forward GRU reads the input sequence s_i as it is ordered from x_{i1} to x_{im} to calculate a sequence of forward hidden states $\overrightarrow{h}_{i1}, \ldots, \overrightarrow{h}_{im}$. Meanwhile, the backward GRU reads the input sequence as it is ordered from x_{im} to x_{i1} to calculate a sequence of forward hidden states $\overleftarrow{h}_{im}, \ldots, \overleftarrow{h}_{i1}$. The computations are listed as follows:

$$x_{it} = W_e w_{it}, \ t \in [1, m] \tag{5}$$

$$\overrightarrow{h}_{it} = \overrightarrow{GRU}(x_{it}), \ t \in [1, m] \tag{6}$$

$$\overleftarrow{h}_{it} = \overleftarrow{GRU}(x_{it}), \ t \in [m, 1] \tag{7}$$

Finally, $h_{it} = \left[\overrightarrow{h}_{it}, \overleftarrow{h}_{it} \right]$ which summarizes the information of the whole sentence s_i centered around w_{it} is obtained.

3. **Word Attention:** Each word in a sentence s_i may not contribute equally to the representation of the meaning of a sentence. Thus, attention mechanism is included to extract and aggregate important words that contribute to the meaning of a sentence as a sentence vector as follows:

$$u_{it} = tanh(W_w h_{it} + b_w) \tag{8}$$

$$\alpha_{it} = \frac{\exp(u_{it}^\top u_w)}{\sum_i \exp(u_{it}^\top u_w)} \tag{9}$$

$$s_i = \sum_t \alpha_{it} h_{it} \tag{10}$$

where u_{it} is a hidden representation of h_{it} and u_w is randomly initialized and jointly learned during the training process.

4. **Sentence Encoder:** Given the sentence vector s_i, bidirectional GRU is applied to encode the sentences as follows:

$$\overrightarrow{h_i} = \overrightarrow{GRU}(s_i),\ i \in [1, n] \tag{11}$$

$$\overleftarrow{h_i} = \overleftarrow{GRU}(s_i),\ i \in [n, 1] \tag{12}$$

Finally, $h_i = \left[\overrightarrow{h_i}, \overleftarrow{h_i} \right]$ which summarizes the neighbour sentences around sentence i is obtained.

5. **Sentence Attention:** Each sentence may not contribute equally to the representation of the classification of a document. Thus, attention mechanism is included to extract and aggregate important sentences that contribute to the classification of a document as a document vector v as follows:

$$u_i = tanh(W_s h_i + b_s) \tag{13}$$

$$\alpha_i = \frac{\exp(u_i^\top u_s)}{\sum_i \exp(u_i^\top u_s)} \tag{14}$$

$$v_i = \sum_t \alpha_i h_i \tag{15}$$

where u_s is randomly initialized and jointly learned during the training process.

6. **Document Classification:** Document vector v can be used as features for document classification as follows:

$$p = \mathsf{softmax}(W_c v + b_c) \tag{16}$$

Finally, negative log likelihood of the correct labels is used as training loss as follows:

$$L = -\sum_d \log p_{dj} \tag{17}$$

where j is the label of document d.

3 Experiments

HAN was applied on six publicly available datasets by Yang et al. [25], and the results of HAN showed better accuracy as compared to the existing methods. These six publicly available datasets include the following:

1. Yelp'13, Yelp'14 and Yelp'15 [23]: Yelp reviews are obtained from the Yelp Dataset Challenge in 2013, 2014 and 2015. Ratings are given from 1 to 5 where higher rating is better.
2. IMDB reviews [6]: User ratings are given from 1 to 10 where higher rating is better.
3. Yahoo answers [26]: The document includes question titles, question contexts and best answers over 10 different classes.
4. Amazon reviews [26]: User ratings are given from 1 to 5 where higher rating is better.

Table 1 presents the summary of all these six datasets being tested using HAN by Yang et al. [25].

Table 1. Data statistics: #s denotes the number of sentences per document, #w denotes the number of words per document, word frequency is the ratio of # document to vocabulary [25]

	Yelp 2013	Yelp 2014	Yelp 2015	IMDB review	Yahoo answer	Amazon review
# Classes	5	5	5	10	10	5
# Documents	335,018	1,125,457	1,569,264	348,415	1,450,000	3,650,000
Average #s	8.9	9.2	9.0	14.0	6.4	4.9
Maximum #s	151	151	151	148	515	99
Average #w	151.6	156.9	151.9	325.6	108.4	91.9
Maximum #w	1184	1199	1199	2802	4002	596
Vocabulary	211,245	476,191	612,636	115,831	1,554,607	1,919,336
Word frequency	1.5859	2.3635	2.5615	3.0080	0.9327	1.9017

From Table 1, we observe the common similarities in the six publicly available datasets tested with HAN by Yang et al. [25]. These six datasets share the following similarities:

1. Each document is written by normal user consisting user's opinion toward certain topic.
2. Each document contains at least 4.9 sentences with 91.9 words in average.
3. Each dataset contains more than 200,000 vocabulary.
4. Each word appears in a dataset 0.9327 to 3.008 times in average.

Thus, experiments are conducted with the aim of answering the following research questions:

- **RQ1:** Do HAN proposed by Yang et al. [25] outperform the state-of-the-art methods in classifying different types of documents, e.g., questions, news article prepared by professional, brief description (snippet) and user review?
- **RQ2:** Do HAN proposed by Yang et al. [25] outperform the state-of-the-art methods in classifying the document which contains lesser training data, smaller vocabulary and/or lesser words?

3.1 Datasets - Selection

To answer **RQ1** and **RQ2**, the following four publicly available datasets are selected:

1. **TREC** [14]: Consists of a set of questions only (without user answers) that can be classified into six classes. These six classes are ABBREVIATION, DESCRIPTION, ENTITY, HUMAN, LOCATION and NUMERIC. One of the samples from the class DESCRIPTION is presented as follows for illustration purpose.

   title What is the oldest profession?

2. **Reuters-21578** [22]: Contains different business and financial news over more than 100 classes. Only eight classes with higher number of document per class are considered in this paper. These eight classes are EARN, ACQ, MONEY-FX, GRAIN, TRADE, INTEREST and SHIP. One of the samples from the class MONEY-FX is presented as follows for illustration purpose.

   ZAMBIA TO RETAIN CURRENCY AUCTION, SAYS KAUNDA Zambia will retain its foreign-exchange auction system despite the suspension of weekly auctions since January 24, President Kenneth Kaunda said.

3. **Amazon** [3]: Product reviewers acquired from Amazon over four different sub-collections, that is, BOOK, DVD, ELECTRONIC and KITCHEN. One of the samples from the class DVD is presented as follows for illustration purpose.

   I saw the scene,where they have Lissa chained to the pool table and gagged in the basement.I didn't understand most of the movie. I bet Kim Possible,Ron Stoppabl,and Rufus can deal with them

4. **Snippets** [20]: Contains word snippets collected from the Google search transactions that can be classified into eight classes. These eight classes are BUSINESS, COMPUTERS, CULTURE-ARTS-ENTERTAINMENT, EDUCATION-SCIENCE, ENGINEERING, HEALTH and SCIENCE. One of the samples from the class HEALTH is presented as follows for illustration purpose.

wikipedia wiki clinic clinic wikipedia encyclopedia clinic outpatient clinic
public facility care ambulatory patients clients

Table 2 presents the summary of all selected four datasets. Different types of documents are included in these four datasets, that is, news article from Reuters, user review from Amazon, short description from Google snippets and question from TREC. The total documents for each selected dataset are at least 28 times smaller than to those datasets being tested by Yang et al. in [25] (see Table 1 for comparison). Similarly, the vocabulary for each selected dataset is at least five times smaller as compared of the datasets presented in Table 1. Thus, each word appears in the four selected datasets 0.2044 to 0.6257 times only in average.

Table 2. Data statistics: #s denotes the number of sentences per document, #w denotes the number of words per document, word frequency is the ratio of # document to vocabulary

	TREC	Reuters	Amazon	Snippets
# Classes	6	8	4	8
# Documents	5,952	7,528	8,000	12,340
Document type	Question	News	Review	Snippet
Average #s	1	6	7	1
Maximum #s	1	68	207	1
Average #w	10	138	128	17
Maximum #w	38	1,322	5,160	38
Vocabulary	9,513	23,582	39,133	29,276
Word frequency	0.6257	0.3192	0.2044	0.4215

3.2 Baseline

The following models are described and are included as baseline for performance comparison.

1. **BOW (binary)** [22]: All documents are pre-processed into bag-of-words vectors. If a word is present in the sentence, then its entry in the vector is 1; otherwise 0. Support vector machine (SVM) method is used to perform text classification.
2. **Centroid** [22]: Documents are projected in the word embedding space as the centroids of their words. Similarity of the documents is then computed using cosine similarity for text classification.
3. **NBSVM** [22]: Wang and Manning [24] combined both Naive Bayes classifier with SVM to achieve remarkable results on several tasks. Rousseau et al. [22] used a combination of both unigrams and bigrams as underlying features.
4. **WMD** [22]: Word Mover's Distance (WMD) is used to compute distances between documents [12]. Rousseau et al. [22] used pre-trained vectors from word2vec (i.e. a two-layer neural networks that are trained to learn linguistic

contexts of words from a large corpus of text) to compute distance between documents. Text classification is done with k-nearest neighbors (KNN) algorithm with the distances between documents. Notice that KNN algorithm classifies an object based on a majority vote of its k neighbors.

5. **CNN** [22]: CNN [13] exploits layer with convolving filters that are applied to local feature. Kim [10] showed that a simple CNN with little hyperparameter tuning and static vectors achieves excellent results for sentence-level classification tasks.

6. **DCNN** [9]: Dynamic k-max pooling is used on top of CNN for the semantic modelling of sentences.

7. **Gaussian** [22]: Short texts are treated as multivariate Gaussian distributions based on the distributed representations of its words. Subsequently, the similarity between two documents is then measured based on the similarity of their distributions for classification.

8. **DMM** [18]: Dirichlet Multinomial Mixture (DMM) model assumes that all documents are generated from a topic. Given the limited content of short texts, this assumption is reasonable.

9. **GPU-DMM** [15]: Inspired by DMM and the generalized Pólya urn (GPU) model, GPU-DMM was proposed by Li et al. [15] to promote the semantically related words under the same topic during the sampling process.

10. **BTM** [4]: Biterm Topic Model (BTM) learns the topics by modeling the generation of word co-occurrence patterns in short texts [4]. Biterm from BTM is an unordered word pair co-occurred from short context.

11. **Bi-RNN + Topic** [7]: Short texts are classified based on abundant representation which utilizes bi-directional recurrent neural network (CNN) with long short term memory (LSTM) and topic model to capture contextual and semantic information.

3.3   Experimental Settings and Results

Different common pre-processing techniques are performed on different selected datasets. These techniques include performing tokenization, removing stop word, removing special character, changing the capitalization of character and selecting pivots with mutual information. For our implemented HAN model, we use pre-trained word embedding vectors from global vectors for word representation (GloVe) to initialize the weight of word embedding layer. Notice that GloVe [19] is an unsupervised learning algorithm for obtaining vector representations for words. Different hyperparameters are set for different datasets as shown in Table 3. Notice that we split each document into a number of sentences denoted as # sentences.

Table 3. Different hyperparameter's settings for different selected datasets

Hyperparameter	TREC	Reuters	Amazon	Snippet
Word embedding dimension	100	200	100	300
GRU dimension	100	100	100	300
# sentences	1	1	10	1
# Training data	5,452	5,485	7,200	10,060
# Testing data	500	2,189	800	2,280

Table 4 shows the comparison of HAN [25] with aforementioned models on TREC, Reuters, Amazon and Snippet datasets in terms of the accuracy of document classification.

Regarding to **RQ1** and **RQ2**, as shown in Table 4, HAN which can pick up important words and sentences in the contextual information is able to out-perform all state-of-the-art models with the improvement of 0.28% to 0.78% in classifying the four public datasets that consists of different types of documents (i.e., question, review, news article and snippets), with smaller size of vocabulary, smaller training data and/or lesser words. This shows that HAN is also suitable for classifying documents with smaller size of vocabulary and lesser words.

Table 4. Comparison of different models for document classification in terms of accuracy

Method	TREC	Reuter	Amazon	Snippet
BoW (binary)	0.9660	0.9571	0.9126	0.6171
Centroid	0.9540	0.9676	0.9311	0.8123
NBSVM	0.9780	0.9712	0.9486	0.6474
WMD	0.9240	0.9502	0.9200	0.7417
CNN	0.9800	0.9707	0.9448	0.8478
DCNN	0.9300	–	–	–
Gaussian	0.9820	0.9712	0.9498	0.8224
DMM	–	–	–	0.8522
GPU-DMM	–	–	–	0.8722
BTM	–	–	–	0.8272
Bi-RNN + topic	0.9400	–	–	0.8636
HAN (this paper)	**0.9860**	**0.9790**	**0.9537**	**0.8750**

4   Visualization of Attention Mechanism

Yang et al. [25] showed that HAN is able to pick up important words and sentences for a user review which consists many words. In this section, we check whether HAN is able to pick up important words for a short question found from the class NUMERIC of TREC dataset. The raw question (without going through pre-processing) randomly selected from TREC is as follows:

dist How far is it from Denver to Aspen?

After going through pre-processing, the question mark is removed as follows:

dist How far is it from Denver to Aspen

Finally, Fig. 2 shows the visualization of attention mechanism for the selected question. Notice that the word with greater red color, the more important the word. This is done by first extracting out the word representation and subsequently coloring each word based on the word representation accordingly. Even though the question is short, HAN is still able to pick up important words that can classify the question as numeric such as "How". On the other hand, the word "is" is not so important for classification.

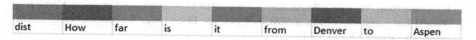

| dist | How | far | is | it | from | Denver | to | Aspen |

Fig. 2. The visualization of attention mechanism for the selected question from TREC (Color figure online)

5 Conclusion

In this paper, our results have demonstrated that HAN is suitable to classify different types of documents (review, question, snippet, and news article) with different sizes. We also showed that HAN is able to pick up important words even for question typed document. However, the improvement of accuracy in classifying short texts cannot be considered as significant. Thus, the future work includes the modification of HAN to further improve the accuracy in classifying both long and short texts.

References

1. Bahdanau, D., Cho, K., Bengio, Y.: Neural Machine Translation by Jointly Learning to Align and Translate. arXiv preprint arXiv: 1409.0473 (2014)
2. Bengio, Y., Ducharme, R., Vincent, P., Jauvin, C.: A neural probabilistic language model. J. Mach. Learn. Res. **3**, 1137–1155 (2003)
3. Blitzer, J., Dredze, M., Pereira, F.: Biographies, bollywood, boom-boxes and blenders: domain adaptation for sentiment classification. In: Carroll, J.A., van den Bosch, A., Zaenen, A. (eds.) Proceedings of the 45th Annual Meeting of the Association for Computational Linguistics (ACL 2007), pp. 440–447. Association for Computational Linguistics, Prague (2007)
4. Cheng, X., Yan, X., Lan, Y., Guo, J.: BTM: topic modeling over short texts. IEEE Trans. Knowl. Data Eng. **26**(12), 2928–2941 (2014)
5. Cho, K., et al.: Learning phrase representations using RNN encoder-decoder for statistical machine translation. In: Moschitti, A., Pang, B., Daelemans, W. (eds.) Proceedings of the 2014 Conference on Empirical Methods in Natural Language Processing (EMNLP 2014), pp. 1724–1734. Association for Computational Linguistics, Doha (2014)
6. Diao, Q., Qiu, M., Wu, C.-Y., Smola, A.J., Jiang, J., Wang, C.: Jointly modeling aspects, ratings and sentiments for movie recommendation (JMARS). In: Macskassy, S.A., Perlich, C., Leskovec, J., Wang, W., Ghani, R. (eds.) Proceedings of the 20th ACM SIGKDD International Conference on Knowledge Discovery and Data Mining (KDD 2014), pp. 193–202. ACM, New York (2014)

7. Gu, Y., et al.: An enhanced short text categorization model with deep abundant representation. World Wide Web **21**(6), 1705–1719 (2018)
8. Hochreiter, S., Schmidhuber, J.: Long short-term memory. Neural Comput. **9**(8), 1735–1780 (1977)
9. Kalchbrenner, N., Grefenstette, E., Blunsom, P.: A convolutional neural network for modelling sentences. In: Toutanova, K., Wu, H. (eds.) Proceedings of the 52nd Annual Meeting of the Association for Computational Linguistics (ACL 2014), pp. 655–665. Association for Computational Linguistics, Baltimore (2014)
10. Kim, Y.: Convolutional neural networks for sentence classification. In: Moschitti, A., Pang, B., Daelemans, W. (eds.) Proceedings of the 2014 Conference on Empirical Methods in Natural Language Processing (EMNLP 2014), pp. 1746–1761. Association for Computational Linguistics, Doha (2014)
11. Androutsopoulos, I., Koutsias, J., Chandrinos, K., Spyropoulos, C.D.: An experimental comparison of Naive Bayesian and keyword-based anti-spam filtering with personal e-mail messages. In: Yannakoudakis, E.J., Belkin, N.J., Ingwersen, P., Leong, M.-K. (eds.) Proceedings of the 23rd International ACM SIGIR Conference on Research and Development in Information Retrieval (SIGIR 2000), pp. 160–167. ACM, Athens (2000)
12. Kusner, M.J., Sun, Y., Kolkin, N.I., Weinberger, K.Q.: From word embeddings to document distances. In: Moschitti, A., Pang, B., Daelemans, W. (eds.) Proceedings of the 32nd International Conference on Machine Learning (ICML 2015), pp. 957–966. Proceedings of Machine Learning Research, Lille (2015)
13. LeChun, Y., Bottou, L., Bengio, Y., Haffner, P.: Gradient-based learning applied to document recognition. Proc. IEEE **86**(11), 2278–2324 (1998)
14. Li, X., Roth, D.: Learning question classifiers. In: Tseng, S.-C., Chen, T.-E. (eds.) Proceedings of the 19th International Conference on Computational Linguistics (COLING 2002), C02-1150. Howard International House and Academia Sinica, Taipei (2002)
15. Li, C., Wang, H., Zhang, Z., Sun, A., Ma, Z.: Topic modeling for short texts with auxiliary word embeddings. In: Perego, R., Sebastiani, F., Aslam, J.A., Ruthven, I., Zobel, J. (eds.) Proceedings of the 39th International ACM SIGIR Conference on Research and Development in Information Retrieval (SIGIR 2016), pp. 165–174. ACM, Pisa (2016)
16. Maas, A.L., Daly, R.E., Pham, P.T., Huang, D., Ng, A.Y., Potts, C.: Learning word vectors for sentiment analysis. In: Lin, D., Matsumoto, Y., Mihalcea, R. (eds.) Proceedings of the 49th Annual Meeting of the Association for Computational Linguistics (ACL 2011), pp. 142–150. Association for Computational Linguistics, Portland (2011)
17. Mnih, A., Kavukcuoglu, K.: Learning word embeddings efficiently with noise-contrastive estimation. In: Burges, C.J.C., Bottou, L., Ghahramani, Z., Wein-berger, K.Q. (eds.) Proceedings of the Advances in Neural Information Processing Systems 26 (NIPS 2013), pp. 2265–2273. Neural Information Processing Systems Foundation, Lake Tahoe (2013)
18. Nigam, K., Mccallum, A.K., Thrun, S., Mitchell, T.: Text classification from labeled and unlabeled documents using EM. Mach. Learn. **39**(2–3), 103–134 (2000)
19. Pennington, J., Socher, R., Manning, C.D.: GloVe: global vectors for word representation. In: Moschitti, A., Pang, B., Daelemans, W. (eds.) Proceedings of the 2014 Conference on Empirical Methods in Natural Language Processing (EMNLP 2014), pp. 1532–1543. Association for Computational Linguistics, Doha (2014)
20. Phan, X.H., Nguyen, M.L., Horiguchi, S.: Learning to classify short and sparse text & web with hidden topics from large-scale data collections. In: Huai, J., et al. (eds.) Proceedings of the 17th International Conference on World Wide Web (WWW 2008), pp. 91–100. ACM, Beijing (2008)

21. Poon, H.-K., Yap, W.-S., Tee, Y.-K., Goi, B.-M., Lee, W.-K.: Document level polarity classification with attention gated recurrent unit. In: Knight, K., Nenkova, A., Rambow, O. (eds.) Proceedings of the 2018 International Conference on Information Networking (ICOIN 2018), pp. 7–12. IEEE, Chiang Mai (2018)

22. Rousseau, F., Vazirgiannis, M., Nikolentzos, G., Meladianos, P., Stavrakas, Y.: Multivariate Gaussian document representation from word embeddings for text categorization. In: Lapata, M., Blunsom, P., Koller, A. (eds.) Proceedings of the 15th Conference of the European Chapter of the Association for Computational Linguistics (EACL 2017), vol. 1432, pp. 450–455. Association for Computational Linguistics, Valencia (2017)

23. Tang, D., Qin, B., Liu, T.: Document modeling with gated recurrent neural network for sentiment classification. In: Màrquez, L., Callison-Burch, C., Su, J., Pighin, D., Marton, Y. (eds.) Proceedings of the 2015 Conference on Empirical Methods in Natural Language Processing (EMNLP 2015), pp. 1422–1432. Association for Computational Linguistics, Lisbon (2015)

24. Wang, S.I., Manning, C.D.: Baselines and bigrams: simple, good sentiment and topic classification. In: Lin, C.-Y., Osborne, M. (eds.) Proceedings of the 50th Annual Meeting of the Association for Computational Linguistics (ACL 2012), pp. 90–94. Association for Computational Linguistics, Jeju Island (2012)

25. Yang, Z., Yang, D., Dyer, C., He, X., Smola, A.J., Hovy, E.H.: Hierarchical attention networks for document classification. In: Knight, K., Nenkova, A., Rambow, O. (eds.) Proceedings of the 2016 Conference of the North American Chapter of the Association for Computational Linguistics: Human Language Technologies (NAACL HLT 2016), pp. 1480–1489. Association for Computational Linguistics, San Diego (2016)

26. Zhang, X., Zhao, J.J., LeCun, Y.: Character-level convolutional networks for text classification. In: Cortes, C.A., Lawrence, N.D., Lee, D.D., Sugiyama, M., Garnett, R. (eds.) Proceedings of the Advances in Neural Information Processing Systems (NIPS 2015), pp. 649–657. Neural Information Processing Systems Foundation, Montreal (2015)

Social Robot Intelligence and Network Consensus

João Silva Sequeira$^{(\boxtimes)}$

Instituto Superior Técnico, Institute for Systems and Robotics, Lisbon, Portugal
joao.silva.sequeira@tecnico.ulisboa.pt

Abstract. Synthetic intelligence models are highly relevant to the fast growing field of social robotics. Robots interacting with humans in the context of a non-lab environment are being upgraded at a fast pace in order to meet social expectations. To foster the integration in human societies, and hence coping with expectations, robots are likely to be endowed with typically human attributes, such as synthetic personality. Such social personality is the result of the interaction of a number of systems, some of high complexity, and can be thought of as a network of dynamic systems. The paper addresses the relations between frameworks consistent with synthetic intelligence models from Psychology, and models of networks with nonsmooth dynamics and consensus problems. The goals are (i) to achieve a framework that can establish bridges between architectures including concepts from social sciences and concepts from nonsmooth dynamic systems, and (ii) to determine the basic properties for such framework. Basic continuity and convexity properties are shown to be at the core of the framework.

Keywords: Social robots · Synthetic intelligence · Consensus problems

1 Introduction

The paper aims at contributing to bridge concepts from social sciences and synthetic intelligence engineering and discuss architectural principles that can be used in the design of social robots, either with an intelligent flavor or aiming simply at basic interactions with humans. It reports ongoing work and builds on (i) theories in synthetic intelligence that support neural-like architectures such as the *Psi* theory (see [2]) and the H-CogAff (see [32]), and (ii) concepts from non-smooth systems, i.e., systems exhibiting some form of discontinuity in some state variables. Robotics is a wide domain that easily leads to non-smooth systems, both in single entity or multiple entities (connected in a network) even when intelligence is not an issue, as recognized, for example, in [10].

Work supported by projects FP7-ICT-9-2011-601033-MOnarCH and FCT [UID/EEA/50009/2013].

Complex systems, such as those found in social robots, regardless of their intelligence attributes, are often developed as a collection of independent, simpler, systems, i.e., a network of systems and all nodes in such networks must behave consistently in order an adequate global behavior is obtained. Such is the case of *Psi* and H-CogAff architectures. Figure 1 shows graphical interpretations of these two architectures. The links between the blocks indicate some form of dependence, exchange of information, and/or temporal synchronization of behavior.

Both architectures in Fig. 1 embed systems normally related to synthetic intelligence in a social environment, namely emotion recognition/generation and their management, and memory management systems. Even though their different nature and theoretical foundations, they both rely on independent blocks and respective interconnections.

The paper is not concerned with argumentative intelligence, i.e., systems like IBM Watson, which are primarily a software system. Instead, we are concerned with systems that must "live" in a social environment, interacting with people, not necessarily showing high argumentative intelligence (though comparative intelligence must be embedded in the systems[1]).

The aforementioned consistency is similar to other problems involving network structures, e.g., modeling of manufacturing processes where manufacturing servers are the nodes in a network (see for instance [5] on deterministic modeling and [11] on stochastic modeling), or having a set of mobile robots moving in formation to complete some assignment, or having a set of software components behaving adequately to some common purpose (see [31]). As pointed in [10], the idea of nodes exchanging information is applicable to a wide range of situations, e.g., computer networks, networks of mobile phones, or air traffic management, human social networks, biology systems, communications networks, transportation systems, among many others (see for instance [26,28] for surveys).

The underlying idea in the paper is to (i) identify models for complex processes, such as those occurring in synthetic intelligence systems, as self-contained units with inputs and outputs, and (ii) the connections between such systems, thus forming a network of dynamic systems. Each of the nodes of the network may have arbitrarily complex dynamics, which often will not be fully known, and hence observers for such dynamics are required.

A node in the network can be thought of as a blackbox that implements some process (this is a specially appealing idea when neural network systems are used - as deep learning strategies are being increasingly used to implement complex input-output maps) coupled with the respective observer system. The observers use any available information from each of the nodes in the network and indicators of the degree of completion of the its assigned task.

A key claim in this network paradigm is that if all nodes can provide performance indicators then by adequately synchronizing relevant events, defined

---

[1] We use here the concepts of comparative and argumentative intelligence in [20], p. 118.

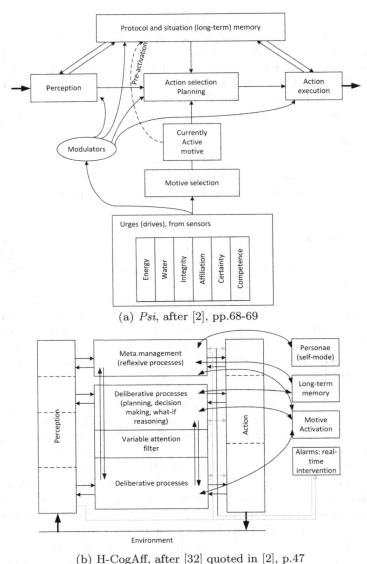

(a) *Psi*, after [2], pp.68-69

(b) H-CogAff, after [32] quoted in [2], p.47

Fig. 1. Architectures for synthetic intelligence systems

after these indicators, it is possible to control the behavior of the whole network and hence achieve what was informally defined above as synthetic intelligence.

This coupling between observers and working systems is a technique used in software development, namely wrapping components with code that publishes information about the internal behavior, making it available to the other components in the system such that they can adjust their own behavior with the status information.

In general, a social robot may be formed by a network with multiple nodes, with subsets related to each of the blocks shown in Fig. 1. The complexity of the mappings implemented by these nodes is clear (as synthetic intelligence has been an elusive goal). Bottom line, a networked robot system may have nodes of very complex dynamics and it may be difficult to foresee the global behavior. Even though each individual node may fully comply with its design requirements, e.g., recognize an emotion from imaging data, and is able to execute its mission successfully, in general it will not be clear that the full network also does it. In a sense, the good properties of each node, when considered individually, are not a sufficient condition for the full network to have them also.

The behavior of the whole network can be related to the adequate sequencing of the individual behaviors of each node, i.e., some form of synchronization of the operation of each node with that of other nodes, such that all nodes reach a consensus on the global behavior. Consensus is thus the key framework to represent the idea of exchanging synchronization information among the nodes.

The paper evolves around the idea that any component in a complex robotic system, namely a social robot, can be viewed as a node in a network and each node has one dynamics that determines the behavior of the node, from a task perspective, and a second dynamics that observes the first one. The task (first) dynamics receives/sends information from/to the network. The observer (second) dynamics senses the task dynamics for adequate state variables and also outputs from other nodes, computes an indication of performance that it outputs to the nodes it is connected to. Often in cyberphysical systems (as in those present in most robots) there is some uncertainty an hence approximate models must be used to develop the observers. Of course, the performance of the overall system is heavily dependent on the quality of the observers.

The paper is organized as follows. Section 2 briefly reviews some examples of social robots and synthetic intelligent paradigms. Section 3 reviews key ideas on consensus problems. Section 4 presents a number of concepts relevant for modeling of network structure using nonsmooth dynamic systems. These are based on the ideas of the two previous sections and develops some arguments to support the claim that systems implementing synthetic intelligence must follow well defined properties from dynamics systems. Section 5 summarizes the main arguments outlined in the paper.

2   Anatomy of a Social Robot

The literature on social robots is extensive. Though most of the examples refer to laboratory-like experiments, in which the environment is controlled, the ideas outlined in this paper still apply.

The social robot Maggie (see, for instance [17] for pictures and details) includes modules for (i) speech recognition and generation, with multimodal fusion and fission mechanisms, (ii) emotion control, (iii) motion behaviors, and (iv) behavioral composition. Figure 2 shows a diagram interpretation of the systems embedded in Maggie. The nodes of the diagram represent functional/behavioral skills. The edges (directionality not represented) indicate a dependency.

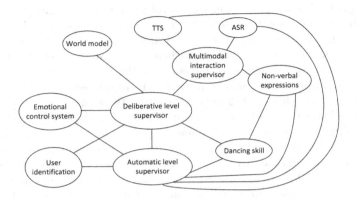

Fig. 2. The Maggie robot systems (adapted from the information in [17])

Social robots for guidance/touring and assistance in public spaces have been described in [19]. Figure 3 shows a diagram similar to the previous one, with the main functionalities/behaviors in the Robovie robot (see [19] for pictures and details).

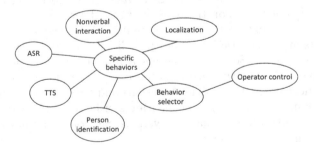

Fig. 3. The Robovie systems (adapted from the information in [19])

Even for such small number of nodes the interaction may be highly complex. In complex cases, such as the Robovie operating in the Osaka Science Museum which can exhibit hundreds of different behaviors (see [19]), the probable behavior of the robot can be anticipated. However, complex interactions among the components may result in unexpected/abnormal behaviors difficult to foresee.

Other well known examples like Pepper and Nao, from Softbank Robotics[2], and the Sanbot, [7], are now in the commercial domain though their social skills, namely in what can be considered synthetic intelligence, are still somewhat limited.

As the above examples show, network structures can be commonly used to model a social robot. The technologies used in some of the nodes may yet require some development, e.g., as those concerning emotion recognition, and the current trends on Artificial Intelligence, namely the growing usage of (deep) neural networks to model complex systems creates difficulties in identifying formal models and hence estimating relevant properties of the full systems.

The Monarch robot, [30], embeds, among others, the following systems (each of them in different degrees of development), (i) Text-to-speech (TTS), (ii) Speech recognition (ASR), (iii) Localization, (iv) Big obstacle detection via Kinect depth images, (v) Liveliness interaction, (vi) People detection using RFID, (vii) People detection using Kinect depth images, (viii) Cellphone based remote interface, (ix) Watchdog to monitor some safety indicators, (x) Robot moving detection, (xi) Robot lost detection, (xii) Adjust the goal of a pre-defined mission. These are just some of the systems a priori assumed as essential for a proper social behavior to be achieved. Therefore, if the network interaction among the systems is adequate the robot will be able to "survive" in a non-lab social environment.

Figure 4 shows a diagram where the edges represent dependencies between the previously referred systems (i.e., nodes). The TTS and ASR nodes are specially interesting as the most promising implementations nowadays available are available as remote services through internet, thus fitting nicely in this network paradigm.

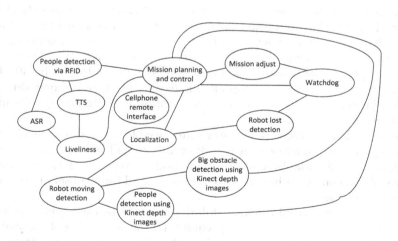

Fig. 4. Interconnections among systems in the MOnarCH social robot

---

[2] https://www.softbankrobotics.com/emea/en.

The information exchanged through the edges in the graph of Fig. 4 can be of multiple different type, e.g., numerical, and categorical, and in general each edge may correspond to multiple physical/virtual connections. The network in Fig. 4 can be expanded such that each node dynamics can be detailed, possibly as another network. The following sections consider generic networks, though the classes of dynamics considered can accommodate a wide enough class of complex systems, such as those in Fig. 4.

3 Consensus Problems

When a subset of entities has to synchronize the evolution of their dynamics we have a consensus problem (CP), e.g., as when having state variables of different entities converging to limit values. CP based decisions are used, explicitly, in multiple situations in everyday life, e.g., when a sequence of intermediate decisions leads must be verified/confirmed before a final decision. Consensus has been referred as a necessary condition for multi-agent coordination (see [29]). For practical purposes, the consensus can be achieved through the convergence of functions of state variables. Each entity (or node of the network) can use the state and consensus information observed from other nodes to control its own dynamics. This is usually called the consensus protocol.

A survey of relevant results in CP is presented in [12]. Networks of agents with double-integrator dynamics and periodic sampling communication protocols have been considered in [18]. When the topology of the network is fully interconnected, nodes have single integrator dynamics and are time invariant, and the consensus protocol is linear, i.e., it just scales the consensus information received from the other agents, a consensus can be reached, both in the continuous and discrete cases (see for instance [27,29]). This was extended to nodes with generic time invariant linear dynamics with bounded uncertainties, [12,13]. Uncertainty due to delays and parametric uncertainty in networks with linear time invariant nodes have been considered in [6] and [21]. A network of nonlinear systems with linear, time varying, protocol is addressed in [40]. Each node in the network with the corresponding feeding protocol is required to have a stable dynamics and there is a zero-sum relation between the weights in the connections input to a node.

Assuming that a consensus indicator state variable in a node can be identified with a degree of completeness of a task (which should not be difficult as, by definition, a state variable reports information characterizing the dynamics), then reaching a consensus in a network means that all (or a subset of) the nodes completed their tasks.

In what concerns social robots, exhibiting social skills (and synthetic intelligence to live in a social environment) requires (i) the decoupling of a global task into the individual tasks to be assigned to each node, and (ii) the synchronization of the individual nodes involved in the task (not necessarily the whole network), and hence it is a CP. For example, exhibiting a specific personality, e.g., neurotic, may require having the robot exhibiting sudden movements, repeating voice interactions on a constant theme (as if showing some nervousness about some social topic). Complex personalities are formed by simpler personality traits (see for instance the Big Five model, [38], which states that the fundamental features of personality can be decomposed in five traits, or the 16PF model, [9], which distinguishes sixteen factors). Though a purely neurotic social robot may not be very interesting, this trait can be used to generate interesting social behaviors. These characteristics require, among others, that (i) the subsystems related to motion be configured with specific parameters, and (ii) movement may occur with voice interaction, hence a need of synchronization with motion systems.

4   Networked Models and Hybrid Systems

This section summarizes concepts in [31] that are also relevant to the synthetic intelligence network representation, that is the focus of this paper. Figure 5 shows a proposal for the architecture of each node in the network. The specific task/action to be executed by the node is represented by the "Task Dynamics" (TD) block. The "Observer Dynamics" (OD) block stands for the observer introduced in Sect. 1, i.e., the system that controls the exchange of information on local performance with the other nodes.

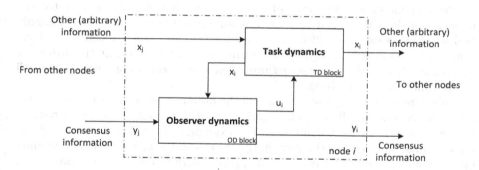

Fig. 5. Node architecture

The information received by the OD block from the TD block is used to determine the consensus information to broadcast to the network. In general the TD will not perfectly known. However, the OD must sense state information, from which the performance estimate is obtained, including this uncertainty.

This node architecture bears inspiration in (i) the dependability consensus building in the context of wireless networks, in [25], namely in what concerns the use of "watchdogs" to detect relevant conditions in the system and use that information to adjust the operation of the network, and (ii) the consensus in cooperative transport systems, [39]. It also has also close connections with concepts from control theory (see [34,41]).

As aforementioned when detailing the node architecture, often the dynamics of a cyberphysical system, as the nodes in the network of Fig. 4, is subject to relevant uncertainties. To make node dynamics more general assume that state variables include both continuous, q_c, and discrete, q_d, variables. In addition, it can also be assumed that each node has both continuous and discrete exogenous inputs. A (significantly) wide class of systems, able to model processes in social robotics subject to uncertainties, is given by a pair of coupled systems of the form

$$\dot{q}_c \in F_c(q_c, q_{d_n}, u_c) \tag{1}$$

$$q_{d_{n+1}} \in F_d(q_{d_n}, q_c, u_d), \tag{2}$$

where F_c and F_d are set-valued maps and u_c, u_d stand for the inputs to the continuous and discrete dynamics, respectively, and n is the discrete time index. For the sake of simplicity the continuous time variable is omitted. This class of systems copes with a wide range of implementation practices, namely those of having a finite state machine supervising the observer dynamics.

Systems of the form (1) embed (i) continuous and discrete dynamics, (ii) bounded uncertainty about system evolution, and (iii) the coupling between both continuous and discrete dynamics. This class of systems fits the definition of hybrid system in [24], p. 4. We argue that this type of systems can model complex network processes, such as those abstracted in Fig. 5. Moreover, we argue that these network processes can be used to model processes as those involved in intelligent activities of social robots. If the OD can monitor the behavior of the TD, even if there is a neural structure (as these are commonly tied to intelligence - see [23]), i.e., as a blackbox that was trained to perform some task and simply integrated as an independent component (regardless of its potentially complex inner structure), then some form of consensus among all components can be obtained.

Solutions for this type of systems can be obtained by making the set-valued maps to account for all admissible variations of the inputs, leading to systems that can be written more compactly. For example, the explicit influence of discrete inputs that control the switching between different continuous dynamics can be removed as (see for instance [33], p. 32), with t and n the continuous and discrete times, respectively,

$$F_c(q_{c_t}, q_{d_n}) \equiv \cup_{u_c} F_c(q_{c_t}, q_{d_n}, u_c)$$
$$F_d(q_{d_n}, q_{c_t}) \equiv \cup_{u_d} F_d(q_{d_n}, q_{c_t}, u_d) \tag{3}$$

yielding,

$$\dot{q}_{c_t} \in F_c(q_{c_t}, q_{d_n}) \tag{4}$$

$$q_{d_{n+1}} \in F_d(q_{d_n}, q_{c_t}) \tag{4b}$$

The existence of solutions for differential inclusion systems, as in the case of systems of the form (4), requires the convexity and upper semi-continuity (USC) (see for instance [1,33,36]) of the composition $F_c \circ F_d^\star$, where $F_d^\star$ stands for a solution map of the discrete dynamics. The same requirement is valid for a smaller class of systems, such as that in [31] in which the discrete inclusion in (4b) is reduced to an inclusion not expressing a temporal evolution - a finite state machine implementing a feedthrough map.

In what concerns the algebraic difference equation (4b), solutions depend on the composition $F_d \circ F_c^\star$, where $F_c^\star$ stands for a solution map of the continuous dynamic, and USC is also required, [15].

The solution maps of convex and USC righthand side inclusions, as in (4), are known to be absolutely continuous, [3]. Absolute continuity amounts to have a bound on the combination of all discontinuities (see for instance [22], p. 24), and hence it is stronger than simply USC. In particular, solutions may be lower semi-continuous (LSC). Also, almost-USC, also known as approximate-USC and a weaker condition than USC (that is a USC map is also almost-USC, see [37], p.163), righthand side inclusions yield USC solution maps, [8], and hence USC is still a key property for the existence of solutions of the system (4), (4b).

Assuming topological domain and co-domain spaces, for set-valued maps, the composition of USC maps is also USC (see Proposition 5.3.9 in [14], p. 95).

Following the ideas outlined in [31], the network model proposed in Sect. 2 can be represented by stacking the individual models of the nodes. Stacking assumes that there are no dimensionality/complexity concerns and the network can simply be represented as a single large, extended, system. In what concerns social robotics, the current examples of architectures suggest that the dimension of the resulting stacking is not significantly large.

Stacking the node dynamics (4), (4b), in the full network preserves the USC property in the global system. The stacking operation of generic maps F_1 and F_2 can be decomposed as $(F_1, 0_n) + (0_m, F_2)$, each f the terms in the sum being a cartesian product between a space of maps and the null elements, 0_m and 0_m, of adequate dimension (n and m).

Under compactness assumption, the cartesian product or upper (lower) semi-continuous maps is also upper (lower) semicontinuous, (see for instance [35], p. 240), and using also Proposition 5.3.9 or Proposition 5.3.11 both in [14], p. 95, USC is ensured for the stacked system, i.e., the whole network is a USC map. Moreover, stacking also preserves convexity (as the cartesian product of convex sets is convex).

Reaching a consensus means reaching an equilibrium in a subset of consensus indicators of the OD (the subset relevant for the assigned mission). Lyapunov generalized theory (see for instance Theorem 1, Chap. 6, in [1]) can be used to assess the existence of equilibria in systems in the form (4). The discrete dynamics (4b) expands the images of (4) in the composition process (preserving USC under adequate conditions as seen above). From a practical perspective, assuming that a convex inner hull can be found for the F_c maps, then each node (and also the full network) generates a solution. This forms a first design principle for the controller at each node, i.e., they must be able to constrain the state evolution to a subset of the accessible set.

When the nodes are stacked, following [1], Definition 1 and Corollary 2, pp. 291–292, if the evolution rate of a Lyapunov function (the contingent derivative) is upper bounded by a negative function converging to 0 then there is an equilibrium, this meaning a consensus. Moreover, this is a globally, asymptotically stable equilibrium (see for instance Theorem 2.10 in [4]). This forms the second principle to design the node controllers.

5 Conclusions

The paper merges concepts commonly used in synthetic intelligence architectures with concepts from the theory of nonsmooth dynamical systems. Known architectures and social robots are used to justify a network of hybrid non-smooth dynamic systems as a generic model for such systems. Systems with discontinuous inputs are naturally to be expected in a networked system of this nature and hence the framework of differential and difference inclusions seems adequate to capture any relevant phenomena (see [16]). Social interaction, as expected from a social robot, is thus identified with the ability of a robot to execute missions that require behavioral synchronization among the dynamic systems in such network. A number of results from the literature are collected and combined to yield basic properties required from the network nodes for the existence of a solution.

An interesting conclusion is that basic topological properties, namely, upper-semicontinuity and convexity, play a key role for the whole network to be able to reach a stationary solution, which is identified with reaching a consensus i.e., the completion of a mission/task. The paper ends with a suggestion of Lyapunov methods for the design of the controller at each node.

Future work will address the switching between missions/tasks. In general, different tasks may have different requirements for some/all of the components in the network, possibly requiring changes in the structure of some/all of the ODs.

References

1. Aubin, J., Cellina, A.: Differential Inclusions. Grundlehren der mathematischen Wissenschaften, vol. 264. Springer, Heidelberg (1984). https://doi.org/10.1007/978-3-642-69512-4
2. Bach, J.: Principles of Synthetic Intelligence. Oxford Series on Cognitive Models and Architectures. Oxford University Press, Oxford (2009)
3. Benaïm, M., Hofbauer, J., Sorin, S.: Stochastic approximations and differential inclusions. SIAM J. Control Optim. **44**(1), 328–348 (2005)
4. Bernuau, E., Efimov, D., Perruquetti, W.: Robustness of homogeneous and locally homogeneous differential inclusions. In: Proceedings of the European Control Conference (ECC), Strasbourg, France, June 2014 (2014)
5. Bogdan, S., Lewis, F., Kovacic, Z., Mireles Jr., J.: Manufacturing Systems Control Design: A Matrix Based Approach. Springer, London (2006). https://doi.org/10.1007/1-84628-334-5
6. Branicky, M., Phillis, S., Zhang, W.: Stability of networked control systems: explicit analysis of delay. In: Proceedings of the American Control Conference, June 2000

7. Brant, T.: Meet Sanbot, the Watson-Powered Droid, Here to Serve. PC Magazine, March 2017. https://www.pcmag.com/news/352776/meet-sanbot-the-watson-powered-droid-here-to-serve. Accessed July 2018

8. Cârjă, O.: Qualitative properties of the solution set of differential inclusions. Technical report, Scientific Report on the implementation of the project PN-II-ID-PCE-2011-3-0154, November 2011–December 2013 (2013)

9. Cattell, R.: The Scientific Analysis of Personality. Penguin Books, Baltimore (1965)

10. Cortés, J.: Discontinuous dynamical systems: a tutorial on solutions, nonsmooth analysis, and stability (2009). https://arxiv.org/pdf/0901.3583.pdf. Accessed Aug 2018

11. Curry, G., Feldman, R.: Manufacturing Systems Modeling and Analysis. Springer, Heidelberg (2009). https://doi.org/10.1007/978-3-540-88763-8

12. Fax, J., Murray, R.: Information flow and cooperative control of vehicle formations. IEEE Trans. Autom. Control 49(9), 1465–1476 (2004)

13. Figueiredo, L., Santana, P., Alves, E., Ishihara, J., Borges, G., Bauchspiess, A.: Robust stability of networked control systems. In: Proceedings of 7th IEEE Conference on Control and Automation, ICCA, December 2009

14. Geletu, A.: Introduction to Topological Spaces and Set-Valued Maps, Lecture Notes. Institute of Mathematics, Department of Operations Research & Stochastics, August 2006

15. Goebel, R., Teel, A.: Solutions to hybrid inclusions via set and graphical convergence with stability theory applications. Automatica 42(4), 573–587 (2006)

16. Goodall, D., Ryan, E.: Feedback controller differential inclusions and stabilization of uncertain dynamical systems. SIAM J. Control Optim. 26(6), 1431–1441 (1988)

17. Gorostiza, J., et al.: Multimodal human-robot interaction framework for a personal robot. In: Proceedings of RO-MAN 2006 (2006)

18. Liu, H., Xie, G., Wang, L.: Necessary and sufficient conditions for solving consensus problems of double-integrator dynamics via sampled control. Int. J. Robust Nonlinear Control 20, 1706–1722 (2009)

19. Kanda, T., Ishiguro, H.: Human-Robot Interaction in Social Robotics. CRC Press, Boca Raton (2013)

20. Kant, I.: Anthropology From a Pragmatic Point of View. Southern Illinois University Press, Carbondale (1978). Translated by Victor Lyle Dowell and Hans H. Rudnick (eds.)

21. Kobayashi, K., Hiraishi, K.: Design of networked control systems using a stochastic switching systems approach. In: Proceedings of IECON 2012–38th Annual Conference on IEEE Industrial Electronics Society (2012)

22. Leine, R., van de Wouw, N.: Stability and Convergence of Mechanical Systems with Unilateral Constraints. Springer, Heidelberg (2008). https://doi.org/10.1007/978-3-540-76975-0

23. Lu, H., Li, Y., Chen, M., Kim, H., Serikawa, S.: Brain Intelligence: Go Beyhond Artificial Intelligence (2018). https://arxiv.org/ftp/arxiv/papers/1706/1706.01040.pdf. Accessed Sept 2018

24. Lunze, J.: What is a Hybrid System? In: Engell, S., Frehse, G., Schnieder, E. (eds.) Modelling, Analysis, and Design of Hybrid Systems, pp. 3–14. Springer, Heidelberg (2003). https://doi.org/10.1007/3-540-45426-8_1

25. Matsuno, Y., Yamamoto, S.: A framework for dependability consensus building and in-operation assurance. J. Wirel. Mob. Netw. 4(1), 118–134 (2012)

26. Namatame, A., Chen, S.: Agent-Based Modeling and Network Dynamics. Oxford University Press, Oxford (2016)

27. Olfati-Saber, R., Alex Fax, J., Murray, R.: Consensus and cooperation in networked multi-agent systems. Proc. IEEE **95**(1), 215–233 (2007)
28. Porter, M., Gleeson, J.: Dynamical Systems on Networks: A Tutorial (2015). arXiv:1403.7663v2 [nlin.AO]
29. Ren, W., Beard, R., Atkins, E.: A survey of consensus problems in multi-agent coordination. In: Proceedings of American Control Conference, Portland OR, USA, 8–10 June 2005 (2005)
30. Sequeira, J., Lima, P., Saffiotti, A., Gonzalez-Pacheco, V., Salichs, M.: MOnarCH: Multi-Robot Cognitive Systems Operating in Hospitals, Karlsruhe, Germany (2013)
31. Sequeira, J.: Dependability in robotics as a consensus problem. In: Proceedings of ROBIO 2017, Macau, PRC, 5–8 December 2017 (2017)
32. Sloman, A., Chrisley, R., Scheutz, M.: The architectural basis of affective states and processes. In: Fellous, J., Arbib, M. (eds.) Who Needs Emotions? The Brain Meets the Robot, pp. 203–244. Oxford University Press, Oxford (2005)
33. Smirnov, G.: Introduction to the Theory of Differential Inclusions. Graduate Studies in Mathematics, vol. 41. American Mathematical Society, Providence (2001)
34. Sontag, E.: Input to state stability: basic concepts and results. In: Nistri, P., Stefani, G. (eds.) Nonlinear and Optimal Control Theory, pp. 163–220. Springer, Heidelberg (2006). https://doi.org/10.1007/978-3-540-77653-6_3
35. Takayama, A.: Mathematical Economics, 2nd edn. Cambridge University Press, Cambridge (1985)
36. Taniguchi, T.: Global existence of solutions of differential inclusions. J. Math. Anal. Appl. **166**, 41–51 (1990)
37. Tarafdar, E., Chowdhury, M.: Topological Methods for Set-Valued Nonlinear Analysis. World Scientific Publishing Co. Pte. Ltd., Singapore (2008)
38. Tupes, E., Christal, R.: Recurrent personality factors based on trait ratings. Technical report ASD-TR-61-97. Lackland Air Force Base, TX, Personnel Laboratory, Air Force Systems Command (1961)
39. Villani, E., Fathollahnejad, N., Pathan, R., Barbosa, R., Karlsson, J.: Reliability analysis of consensus in cooperative transport systems. In: Proceedings of the Workshop ASCoMS (Architecting Safety in Collaborative Mobile Systems) of the 32nd International Conference on Computer Safety, Reliability, and Security - SAFECOMP 2013, Toulouse, France (2013)
40. Wu, C.: Control of networks of coupled dynamical systems. In: Kocarev, L. (ed.) Consensus and Synchronization in Complex Networks, pp. 23–50. Springer, Heidelberg (2013). https://doi.org/10.1007/978-3-642-33359-0_2
41. Wu, C.: On control of networks of dynamical systems. In: Proceedings of 2010 IEEE International Symposium on Circuits and Systems (ISCAS), Paris, 30 May–2 June, pp. 3785–3788 (2010)

UDLR Convolutional Network for Adaptive Image Denoiser

Sungmin Cha and Taesup Moon[✉]

Department of Electrical and Computer Engineering,
Sungkyunkwan University (SKKU), Suwon, South Korea
{csm9493,tsmoon}@skku.edu

Abstract. We propose a new convolutional network architecture called as UDLR Convolutional Network for improving the recently proposed Neural Adaptive Image DEnoiser (NAIDE). More specifically, we develop UDLR filters that meet the conditional independence constraint of NAIDE. By using the UDLR network, we could achieve a denoising result that significantly outperforms the state-of-the-art CNN-based methods on a standard benchmark dataset.

1 Introduction

Image denoising is one of the oldest research topics in image processing. Over the past few decades, many algorithms were proposed and showed good results, e.g., BM3D [3]. However, in recent years, the results of the CNN-based methods were overwhelming compared to the previous methods. Especially, DnCNN [6], RED [4] and Memnet [5] showed the state-of-the-art denoising results. On the other hand, there is a lack of adaptivity in those CNN-based methods, so a new adaptive denoiser using a neural network was recently proposed in [7]. This algorithm showed the possibility of the adaptivity, but the resulting performance was lower than the previous CNN methods. The most critical reason why NAIDE [7] cannot achieve the state-of-the-art denoising result is NAIDE uses only fully connected layers in the neural network model. The CNN-based model can take a local feature of the input image during estimating and training process, while NAIDE cannot. Also, NAIDE requires more parameters than the CNN-based model, and it causes overfitting during the fine-tuning. However, we cannot directly adapt CNN to NAIDE because a naively stacked CNN model breaks the conditional independence condition between a noisy pixel and its contexts, which is an integral condition for NAIDE's adaptivity.

In this paper, we propose a new convolutional network, UDLR convolutional network, that overcomes a limitation mentioned above. To maintain the conditional independence in NAIDE setting, we implement four different convolution layer U (up), D (down), L (left) and R (right). By using UDLR convolutional network, we can achieve the state-of-the-art denoising results in the benchmark dataset.

The architecture of this paper is as follow. First, Sect. 2 reminds problem settings proposed in NAIDE. Section 3 introduces our UDLR convolutional network. In Sect. 4, experimental results are given for the benchmark dataset. Finally, several concluding remarks and future works are given in Sect. 5.

© Springer Nature Singapore Pte Ltd. 2019
J.-H. Kim et al. (Eds.): RiTA 2018, CCIS 1015, pp. 55–61, 2019.
https://doi.org/10.1007/978-981-13-7780-8_5

2 Problem Settings

We follow the same problem setting as [7].

Clean and Noisy Image. We denote $x^{n \times n}$ is the clean grayscale image, and each pixel $x_i \in [0, 255]$. Each pixel is corrupted by an additive noise to result in noisy pixel Z_i, i.e., $Z_i = x_i + N_i \left(E(N_i) = 0, E(N_i^2) = \sigma^2 \right)$.

Estimated Loss Function. For the adaptive image denoising, we derive *Estimated Loss Function* from the true MSE, $\Lambda(x, \hat{X}(Z))$, in [7]:

> *Suppose $Z = x + N$ with $E(N) = 0$ and $E(N^2) = \sigma^2$, and consider*
> *a mapping of form $\hat{X}(Z) = aZ + b$. Then,*
> $$L(Z, (a, b); \sigma^2) = (Z - (aZ + b))^2 + 2a\sigma^2$$
> *is an unbiased estimated of $E\Lambda(x, \hat{X}(Z)) + \sigma^2$*

Affine Denoiser. We define our affine denoiser as $\hat{X}_i(Z^{n \times n}) = a\left(w, C_{k \times k}^{\backslash i}\right) \cdot Z_i + b\left(w, C_{k \times k}^{\backslash i}\right)$. $\hat{X}_i$ is a reconstruction of Z_i and $\left(a\left(w, C_{k \times k}^{\backslash i}\right), b\left(w, C_{k \times k}^{\backslash i}\right)\right)$ are output of our neural network model when fed a context of noisy image centered around but without Z_i to the model.

Two Steps for Neural AIDE. The first step is supervised training. We minimize the following *Supervised Loss Function* by using collected abundant clean and noisy image pairs $(\tilde{x}, \tilde{Z})$:

$$\frac{1}{N} \sum_{i=1}^{N} \left(\tilde{x} - \left(a\left(w, \tilde{C}_{k \times k}^{\backslash i}\right) \cdot \tilde{Z}_i + b\left(w, \tilde{C}_{k \times k}^{\backslash i}\right) \right) \right)^2$$

Given a weight $\tilde{w}$ trained by *Supervised Loss Function*, the second step is adaptive training with given noisy image by minimizing *Estimated Loss Function*. Because we define the affine denoiser as $\hat{X}_i(Z^{n \times n}) = a\left(w, C_{k \times k}^{\backslash i}\right) \cdot Z_i + b\left(w, C_{k \times k}^{\backslash i}\right)$, given $C_{k \times k}^{\backslash i}$, we can show that $L\left(Z_i, \left(a\left(w, C_{k \times k}^{\backslash i}\right), b\left(w, C_{k \times k}^{\backslash i}\right)\right); \sigma^2\right)$ is an unbiased estimate of $E_{Z_i}\left(\Lambda(x_i, \hat{X}_i(Z^{n \times n})) | C_{k \times k}^{\backslash i}\right) + \sigma^2$. Therefore, the final *Estimated Loss Function* as follow:

$$\frac{1}{n} \sum_{i=1}^{n^2} L\left(Z_i, \left(a\left(\tilde{w}, C_{k \times k}^{\backslash i}\right), b\left(\tilde{w}, C_{k \times k}^{\backslash i}\right)\right); \sigma^2\right)$$

After two steps, we get a weight w^* for. As a result, we can reconstruct $\hat{X}_i$ by using w^* and this formula:

$$\hat{X}_{i,\,Neural\,AIDE}(Z^{nxn}) = a(w^*, C_{kxk}^{\backslash i}) \cdot \tilde{Z}_i + b(w^*, C_{kxk}^{\backslash i})$$

3 UDLR Convolutional Network

Neural AIDE [7] got a competitive denoising results compared to other denoising methods. However, the final PSNR on the benchmark datasets is slightly worse than the CNN based model, for example, DnCNN-S [6]. We think that this difference of the performance comes from the power of the Convolutional Neural Network (CNN) used in [6], because in [7], we just used a simple Fully Connected Neural Network (FCNN).

In the course of developing a new convolutional layer for our method, we find out a constraint that the value at the i-th location in any feature maps at any layer should not depend on z_i and have to use only 1×1 convolutional layer for the output layer. This is because to maintain the independence between $\left(a\left(w, C_{kxk}^{-i}\right), b\left(w, C_{kxk}^{-i}\right)\right)$ and z_i.

To meet this constraint, we devise UDLR filters in Fig. 1. Each UDLR filter has an inherent mask denoted as 0. By stacking these UDLR filters, we can get receptive fields in each direction while maintaining the constraint. However, the denoising result of a model that only stacks UDLR filters is not competitive because the model composed of masked convolution layers cannot minimize a training loss in the process of supervised training. To solve this problem, we train some different models to find the best model.

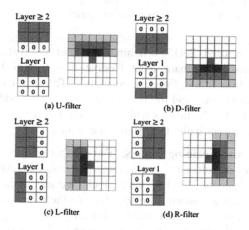

Fig. 1. UDLR filters

From the process of model search, we could find the best model depicted in Fig. 2. As mentioned above, we stacked 22 UDLR Layers, and then averaged the outputs of UDLR layers in each layer level to get the whole receptive field centered around z_i.

We also add 1×1 residual block after an average layer. As a result of model search, we use the receptive field of 45×45.

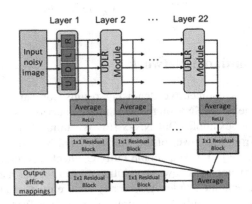

Fig. 2. UDLR convolutional network

4   Experimental Results

Training Data. We collected a total of 3000 images from the Berkeley Segmentation Dataset (BSD) [1] and Pascal VOC 2012 Dataset [2]. Then, we cropped 128 50×50 patches from each image. As a result, we used 128×3000 training images for the supervised learning. During the supervised learning, we used noise augmentation to generate different noise realizations from clean patches size of mini-batch in each epoch.

Evaluation Data. We used Set13 benchmark dataset to evaluate our and other methods. Set 13 consist of [Barbara, Boat, Couple, F.print, Hill, Lena, Man, C.man, House, Peppers] and [Flintstone, Einstein, Shannon]. The first 10 images are widely used images for the evaluation and the remaining 3 images are additionally added for evaluating an adaptivity for the images with very different textures. All evaluation data are corrupted by an additive white Gaussian noise with $\sigma = 30$.

Model Architecture and Training Details. The number of filters in all UDLR layer and 1×1 residual block before the last average layer was 64. After the last average layer, we used 128 filters for 1×1 residual block. We used Adam optimizer for the training and set the initial learning rate to 0.001 in the supervised learning and to 0.0001 in the fine-tuning. Also, we used a learning rate decay in the supervised learning. The size of mini-batch is 64 in the supervised learning and 1 in the fine-tuning.

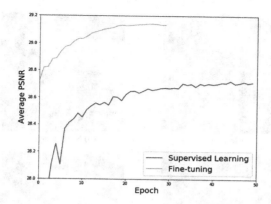

Fig. 3. The graph of supervised learning and fine-tuning

Supervised Training and Fine-Tuning. Figure 3 shows graphs of the supervised training and fine-tuning process in Set13. From these graphs, we can check that our UDLR Convolutional Network can learn and estimate $\left(a\left(w, C_{kxk}^{-i}\right), b\left(w, C_{kxk}^{-i}\right)\right)$ for denoising images. Especially, we also see that the fine-tuning process is effective because it improves PSNR (Peak signal-to-noise ratio) and SSIM (Structural Similarity) result of the supervised learning.

Table 1. The average PSNR (dB) on Set13

	BM3D	RED	DnCNN-S	Memnet	UDLR$_S$	UDLR$_{S+FT}$
PSNR	28.56	28.91	28.64	28.83	28.73	**29.14**
SSIM	0.8491	**0.8605**	0.8537	0.8593	0.8482	0.8547

Denoising Results on Set13. Table 1 summarize the average PSNR on Set 13. We compare our methods with BM3D [3], RED [4], DnCNN-S [6] and Memnet [5] as baselines. From this table, we can see that our method, $UDLR_{S+FT}$, surpasses the average PSNR of other methods with a difference of at least 0.23 dB. Especially, we also check that the average PSNR of $UDLR_S$, which only uses the supervised model to denoise a noisy image without the fine-tuning, is higher than that of $DnCNN-S$. This result means our new proposed model, UDLR Convolutional Network, is very effective for image denoising.

Visualization. We visualize the Barbara image for different cases: clean, noisy, DnCNN, MemNet, UDLR-S (Supervised) and UDLR-S+FT (supervised + Fine-tuning) in Fig. 4. Especially, we focus on the striped area on the table, because it is generally known that the repeated pattern of this area is hard to denoise by using CNN based methods. Comparing with the Noisy image, UDLR-S denoises quite well, but the striped area is not clearer than that of the Clean image; we observe the same phenomenon for DnCNN and MemNet. However, we note that UDLR-S+FT improves a clarity of the striped patterns. This result emphasizes adaptivity of our UDLR Convolutional Layer and shows the improvement over NAIDE.

(a) Clean image

(b) Noisy image

(c) DnCNN
(PSNR : 28.93)

(d) MemNet
(PSNR : 29.11)

(e) UDLR-S
(PSNR : 29.08)

(f) UDLR-S+FT
(PSNR : 29.46)

Fig. 4. The visualization results on Barbara

5 Conclusion

In this paper, we proposed a new convolutional layer, UDLR Convolutional Network, for adaptive image denoising. To maintain the independence condition identified in NAIDE paper, we implemented four different convolution layers: U, D, L and R. Using these layers, we designed ULDR Convolutional Network and got a strong denoising result compared to other state-of-the-art baselines from experiments. Also, we could check that the adaptive image denoising is effective, and a pattern that is hard to denoise when use DNN models is also denoised well by using an adaptivity. For future work, we will focus on two things; implement more sophisticated CNN layer and find out a novel way for the fine-tuning step in order to speed-up the adaptive denoising process.

References

1. Martin, D., Fowlkes, C., Tal, D., Malik, J.: A database of human segmented natural images and its application to evaluating segmentation algorithms and measuring ecological statistics. In: International Conference on Computer Vision (ICCV) (2001)
2. Everingham, M., Van Gool, L., Williams, C.K.I., Winn, J., Zisserman, A.: The PASCAL visual object classes challenge 2012 (VOC2012) results. http://www.pascal-network.org/challenges/VOC/voc2012/workshop/index.html
3. Dabov, K., Foi, A., Katkovnik, V., Egiazarian, K.: Image denoising by sparse 3-D transform domain collaborative filtering. IEEE Trans. Image Process. 16(8), 2080–2095 (2007)
4. Mao, X., Shen, C., Yang, Y.-B.: Image restoration using very deep convolutional encoder decoder networks with symmetric skip connections. In: Neural Information Processing Systems (NIPS) (2016)
5. Tai, Y., Yang, J., Liu, X., Xu, C.: MemNet: a persistent memory network for image restoration. In: ICCV (2017)
6. Zhang, K., Zuo, W., Chen, Y., Meng, D., Zhang, L.: Beyond a Gaussian denoiser: residual learning of deep CNN for image denoising. IEEE Trans. Image Process. 26(7), 3142–3155 (2017)
7. Cha, S., Moon, T.: Neural adaptive image denoiser. In: IEEE ICASSP (2018)

Robust Feedback Controller Design Using Cuckoo Search Optimization to Maximize Stability Radius

Mahmud Iwan Solihin[1](✉) and Rini Akmeliawati[2]

[1] Faculty of Engineering, UCSI University, 56000 Kuala Lumpur, Malaysia
mahmudis@ucsiuniversity.edu.my
[2] School of Mechanical Engineering, University of Adelaide,
Adelaide, Australia

Abstract. A robust feedback controller is designed to maximize complex stability radius via single objective constrained optimization using Cuckoo Search Optimization (CSO) in this paper. A set robust feedback controller gains is optimized based on plant's linear model having structured parametric uncertainty, i.e. two mass benchmark system. A wedge region is assigned as the optimization constraint to specify the desired closed-loop poles location which is directly related to desired time-domain response. The simulation results show that the robustness performance is achieved in the presence of parameter variations of the plant. In addition, the feedback controller optimized by CSO performs slightly better than that optimized by differential evolution algorithm previously designed.

1 Introduction

Robustness is an important issue for any control system design. A successfully designed control system should be always able to maintain stability and performance level despite of uncertainties in the system dynamics including parameter variations of the plant. Robustness, stability and control performance are therefore important aspects in control and robotic applications (Solihin et al. 2011, Tang et al. 2014).

Conventionally, H_∞ optimization approach and the μ-synthesis/analysis method are well developed and elegant for robust control design (Gu et al. 2005). They provide systematic design procedures of robust controllers for linear systems. However, the mathematics behind the theory is not trivial and is not straightforward to formulate a practical design problem into the design framework. It is also often that these conventional robust controller designs are followed by a lengthy tuning of weighting functions.

In this paper, an alternative technique of robust feedback control design via constrained optimization using Cuckoo Search Optimization (CSO) is proposed. CSO is a meta-heuristic optimization algorithm which is introduced based on inspiration from

J.-H. Kim et al. (Eds.): RiTA 2018, CCIS 1015, pp. 62–75, 2019.
https://doi.org/10.1007/978-981-13-7780-8_6

the obligate brood parasitism of some cuckoo species (Yang and Deb 2010). CSO is inspired by some species of a bird family called cuckoo because of their special lifestyle and aggressive breeding strategy. CSO has been demonstrated the efficiency to quickly converge in global optimization problems (Zabihi Samani and Amini 2015). CSO has been utilized newly as a formidable optimization algorithm in engineering problems including control engineering (Sethi et al. 2015; Fatihu Hamza et al. 2017; Jin et al. 2015, Wang et al. 2017; Singh et al. 2016). Rapid convergence, and simplicity in determining algorithm parameters are some merits of CSO (Balochian and Ebrahimi 2013; Wang et al. 2017).

A number of works have proposed modern optimizations, such as using GA (genetic algorithm), DE (Differential Evolution), PSO (particle swarm optimization) or other metaheuristics algorithms in order to overcome the complications in the conventional robust control design (Tijani et al. 2011; Solihin et al. 2014; Feyel 2017).

In particular, CSO has recently received much attention to be applied for controller parameters tuning via optimization (Barbosa and Jesus 2015; Sethi et al. 2015, Kishnani et al. 2014; Khafaji and Darus 2014). However, the controller parameters tuning using CSO in robust control framework has not received much attention thus far. Therefore, this work combines the advantages of modern meta-heuristics optimization algorithm especially CSO with robust control theory. The optimization is performed in a single objective mode instead of multi-objectives. This simplifies the formulation of the optimization.

To deal with the plant's parameter uncertainty the complex stability radius as a tool of measuring system robustness is used. In addition, the desired response is automatically defined by assigning a regional closed loop poles placement. This region will be incorporated in the CSO-based optimization as a constraint. In other word, the proposed controller design technique is searching for a set of robust feedback controller gains such that the closed-loop system would have maximum complex stability radius. Maximizing complex stability radius is therefore taken as the objective of the optimization.

At the end of the work, the simulation results of the proposed robust control design for two-mass system is presented. This two-mass system is commonly known as a benchmark problem for robust control design (Meza et al. 2017; Wie and Bernstein 1992).

2 Problem Formulation

2.1 State Feedback Controller

Consider a plant model of linear time-invariant continuous-time system:

$$\dot{x}(t) = Ax(t) + Bu(t)$$
$$y(t) = Cx(t) \tag{1}$$

with $A \in \mathbb{R}^{n \times n}$, $B \in \mathbb{R}^{n \times m}$, $C \in \mathbb{R}^{l \times m}$, $x(t) \in \mathbb{R}^n$, $u(t) \in \mathbb{R}^m$, and $y(t) \in \mathbb{R}^p$ are state matrix, input matrix, output matrix, state vector, control input and output vectors, respectively. It is assumed that the system given by Eq. (1) is completely state controllable and all state variables are available for feedback. One can use state feedback controller with feed-forward integral gain as shown in Fig. 1. The control signal (u) is given by a linear control law:

$$u(t) = -kx(t) + k_i\xi(t) \tag{2}$$

where $k = [k_1, k_2, k_3, \ldots k_n]$ is the state feedback gain, k_i is integral feedforward gain and ξ is output of the integrator. The controller gains for the system in Fig. 1 consists of the feedback gain k and integral feedforward gain k_i, which can be computed using some conventional techniques such as pole placement or optimal control method which is also known as linear quadratic regulator (LQR). However, these conventional techniques do not consider plant's parametric uncertainty explicitly.

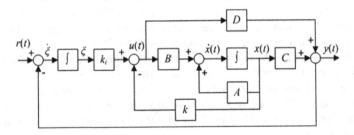

Fig. 1. State feedback controller with feed-forward integral gain

In feedback controller design in Fig. 1, the main objective is to locate the closed-loop poles into a specific region such that the time-domain performance is satisfactory. In addition, the obtained feedback system is also robust to parameter variation of the plant. Therefore, it is naturally a bi-objective problem.

To simplify the design process, this problem is transformed into a single objective constrained optimization. In this work, a single objective constrained optimization using CSO is employed to find a set of robust controller gains ($K = [k\ k_i]$) such that closed-loop system would have maximum stability radius (explained in Sect. 2.2). Plant's parametric uncertainty is automatically handled with the use of stability radius. In addition, a wedge region for closed-loop poles is incorporated as optimization constraint to allow the designers to specify the desired time-domain control performance. For efficiency of the constrained optimization, a dynamic constraint handling technique (explained in Sect. 3.3) is adopted instead of common constraint handling technique such as penalty function approach.

2.2 Stability Radius

In this section, a tool of measuring system robustness called stability radius is presented (Hinrichsen and Pritchard 1986). Stability radius is the maximum distance to instability. Equivalently, a system with larger stability radius implies that the system can tolerate more perturbations. In general, stability radius is classified as complex stability radius and real stability radius. Compared to real stability radius, complex stability radius can handle a wider class of perturbations including nonlinear, linear-time-varying, nonlinear-time-varying and nonlinear-time-varying-and-dynamics perturbations (Byrns and Calise 1992). For this reason, complex stability radius is used as a measure of robustness for the feedback system.

The definition of complex stability radius is given here. Let $\mathbb{C}$ denote the set of complex numbers. $\mathbb{C}_- = \{z \in \mathbb{C}|Real(z) < 0\}$ and $\mathbb{C}_+ = \mathbb{C}\backslash\mathbb{C}_-$; is the closed right half plane. Consider a nominal system in the form:

$$\dot{x}(t) = Ax(t) \tag{3}$$

$A(t)$ is assumed to be stable. The perturbed open-loop system is assumed as:

$$\dot{x}(t) = (A(t) + E\Delta(t)H)x(t) \tag{4}$$

where $\Delta(t)$ is a bounded time-varying linear perturbation. E and H are scale matrices that define the structure of the perturbations. The perturbation matrix itself is unknown. The stability radius of (4) is defined as the smallest norm of Δ for which there exists a Δ that destabilizes (3) for the given perturbation structure (E, H).

For the controlled perturbed system in the form (3), let:

$$G(s) = H(sI - A)^{-1}E \tag{5}$$

be the "transfer matrix" associated with (A, E, H), then the complex stability radius is defined by the following definition.

Definition 1: The complex stability radius, r_c:

$$r_c(A, E, H, \mathbb{C}_+) = \left[\begin{matrix} max \\ s \in \partial\mathbb{C}_+ \end{matrix} \|G(s)\| \right]^{-1} \tag{6}$$

where is the boundary of $\mathbb{C}_+$. In other words, a maximum r_c can be achieved by minimizing the H_∞ norm of the "transfer matrix" G (Akmeliawati and Tan n.d.).

Proposition 1: Using Definition 1, the complex stability radius of the feedback system as shown in Fig. 1 is given as:

$$r_c(\tilde{A}, \hat{E}, \hat{H}, \mathbb{C}_+) = \left[\begin{matrix} max \\ s \in \partial\mathbb{C}_+ \end{matrix} \|\hat{H}(sI - \tilde{A})^{-1}\hat{E}\| \right]^{-1} \tag{7}$$

where $\tilde{A} = \hat{A} - \hat{B}K$, $\hat{A}$ and $\hat{B}$ are given by the following equations:

$$\hat{A} = \begin{bmatrix} A & 0 \\ -C & 0 \end{bmatrix}, \qquad \hat{B} = \begin{bmatrix} B \\ 0 \end{bmatrix}.$$

For the structure of perturbation given by $\hat{E}$ and $\hat{H}$, a robust control system can be obtained by maximizing r_c described by Eq. (7). Therefore, a suitable controller gain K can be optimized by min-max optimization algorithms.

3 Control Design Using Cuckoo Search

3.1 Brief Overview of Cuckoo Search

Since the first introduction of Cuckoo Search (CS) in 2009 (Yang and Deb 2010), the literature of this algorithm has exploded. Researchers tested this algorithm on some well-known benchmark functions and compared with PSO and GA, and it was found that cuckoo search achieved better results than the results by PSO and GA. Since then, the original developers of this algorithm and many researchers have also applied this algorithm to engineering optimization, where Cuckoo search also showed promising results (Fister et al. 2014).

The Pseudo-code of CS is as follows:

Begin
Objective function $f(x)$, $x = (x_1, x_2, \ldots, x_d)^T$;
Generate initial population of N_p host nests $x_i (i = 1, 2, \ldots n)$
While ($t<$ Max Generation) or (stop criterion)
    Get a cuckoo randomly by Levy Flights
    Evaluate its fitness F_i
    Choose a nest among n (say j) randomly
    If ($F_i > F_j$) Replace j by the new solution;
    End If
A fraction (p_a:probability of discovery) of worse nests is abandoned and new ones are built;
Keep the best solutions (or nest with quality solutions)
Rank the solution and find the solution and find the current best
End while
    Post process results and visualization
End Begin

The interesting thing of CS as compared to other optimization algorithms is probably that CS only needs one parameter, i.e. $p_a \epsilon [0, 1]$, in addition to number of population (N_p) which is common all metaheuristics optimization.

3.2 Constrained Optimization

The objective of the optimization is to maximize the complex stability radius (r_c), however in this work the r_c is converted into minimization mode by putting negative sign. Based on our approach, the searching procedure of the robust controller gains using constrained optimization can be formulated as follows (Table 1).

Table 1. Constrained optimization

Minimize:
$f(X) = -r_c(X)$
Subject to constraint:
$\lambda_n(X) \in \psi$ **for n=1,2,...**
and boundary constraint:
$X \in [l_b, u_b]$

where $X = K = (k_1, k_2, \ldots, k_n, k_i)$ is the vector solutions such that $X \in S \subseteq R^{n+1}$. s is the search space, and $F \subseteq S$ is the feasible region or the region of S for which the constraint is satisfied. The constraint here is the closed loop poles region; in the feasible region, the controller gains are found such that the closed loop poles (λ) lie within a wedge region (ψ) of a complex plane as given in Fig. 2. The wedge region can be specified by two parameters θ and ρ which are related to desired transient response characteristics i.e., damping ratio (ζ) and settling time (t_s).

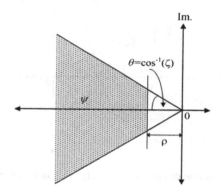

Fig. 2. A wedge region in complex plane for closed loop poles placement

3.3 Constraint Handling

An efficient and adequate constraint-handling technique is a key element in the design of stochastic algorithms to solve complex optimization problems. Although the use of penalty functions is the most common technique for constraint-handling, there are a lot of different approaches for dealing with constraints (Coello Coello 2002).

Instead of using penalty approach, where the optimizer seemed to be inefficient (high iterations), a Dynamic Objective Constraint Handling Method (DOCHM) (Lu and Chen 2006) is adopted here in order to improve the efficiency. By defining the distance function $F(X)$, DOHCM converts the original problem into bi-objective optimization problem $min(F(X), f(X))$, where $F(X)$ is treated as the first objective function and $f(x)$ is the second (main) one.

The auxiliary distance function $F(X)$ will be merely used to determine whether or not an individual (candidate of solution) is within the feasible region and how close a particle is to the feasible region. If an individual lies outside the feasible region (at least an eigenvalue lies outside the wedge region), the individual will take $F(X)$ as its optimization objective. Otherwise, the individual will instead optimize the real objective function $f(X)$. During the optimization process if an individual leaves the feasible region, it will once again optimize $F(X)$. Therefore, the optimizer has the ability to dynamically drive the individuals into the feasible region.

The procedure of the DOCHM applied to the eigenvalue assignment in the wedge region is illustrated in the following pseudo-code (Table 2). Referring to Fig. 3 let d_n is an outer distance of an eigenvalue (λ_n) to the wedge region. It is noted that if an eigenvalue lies within the wedge region, $d_n = 0$. $F(X)$ is defined by:

$$F(X) = \sum_{i=1}^{n+1} \max(0, d_n(\lambda_n(X)))$$ (8)

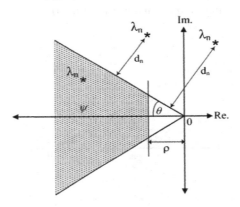

Fig. 3. Eigenvalue distance to the wedge region in complex plane

Table 2. Pseudo-code for the constraint handling

If	$F(X) = 0$
	$f(X) = -r_c(X)$
Else	
	$f(X) = F(X)$
End	

Furthermore, in this study the number of function evaluation (FE) in feasible region is used as stopping criterion, i.e. to terminate the optimization run. This FE counts for how many times the objective function has been evaluated in the optimization loop and the constraint condition are satisfied, i.e. evaluation in feasible region.

4 Case Study: Two-Mass System

In this section, an illustrative example of the application of the proposed method to two-mass system is presented. This system has been used as benchmark problem for robust control design (Wie and Bernstein 1992). Consider the two-mass system shown in the Fig. 4. A control force (u) acts on body 1 and the position of body 2 is measured. Both masses are equal to one unit ($m1 = m2 = 1$) and the spring constant is assumed to be in the range $0.5 \le k \le 2$. The system can be represented as:

$$\begin{bmatrix} \dot{x}_1 \\ \dot{x}_2 \\ \dot{x}_3 \\ \dot{x}_4 \end{bmatrix} = \begin{bmatrix} 0 & 0 & 1 & 0 \\ 0 & 0 & 0 & 1 \\ -k & k & 0 & 0 \\ k & -k & 0 & 0 \end{bmatrix} \begin{bmatrix} x_1 \\ x_2 \\ x_3 \\ x_4 \end{bmatrix} + \begin{bmatrix} 0 \\ 0 \\ 0 \\ 1 \end{bmatrix} u \tag{9}$$

where:

x_1: position of mass-1
x_2: position of mass-2
x_3: velocity of mass-1
x_4: velocity of mass-2

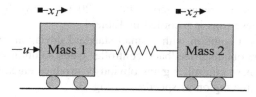

Fig. 4. Two-mass and spring system

The plant uncertainty is due to variations of the spring constant where the nominal value is selected for the worst case of $k = 0.5$. Therefore uncertainties appear in the rows 3–4 and the columns 1–2 of the state matrix. The scale matrices as the perturbation structure for the closed loop system are $\hat{E}$ and $\hat{H}$ whose diagonal elements in rows 3–4 of $\hat{E}$ and in columns 2–3 of $\hat{H}$ are respectively equal to 1.

$$\hat{E} = \begin{bmatrix} 0 & 0 & 0 & 0 & 0 \\ 0 & 0 & 0 & 0 & 0 \\ 0 & 0 & 1 & 0 & 0 \\ 0 & 0 & 0 & 1 & 0 \\ 0 & 0 & 0 & 0 & 0 \end{bmatrix} \qquad \hat{H} = \begin{bmatrix} 1 & 0 & 0 & 0 & 0 \\ 0 & 1 & 0 & 0 & 0 \\ 0 & 0 & 0 & 0 & 0 \\ 0 & 0 & 0 & 0 & 0 \\ 0 & 0 & 0 & 0 & 0 \end{bmatrix}$$

The next is to choose the parameters of the wedge region (Fig. 2) to locate the closed loop poles. The damping ratio is usually set to $\zeta = 0.7$ to produce sufficient overshoot damping in the response. The transient margin (ρ) is specified according to the desired speed of the response. This is problem-dependent parameter and the value of $\rho = 1$ is set. In addition, the main CSO-based optimization parameters are listed in Table 3. A large number of simulation experiments prove that when the bird's nest size (N_p) between 15–40 and the discovery probability of $p_a = 0.25$, it can solve many optimization problems (Wang et al. 2015).

Table 3. CSO-based optimization parameters

Dimension of the problem	D	5
Population of host nest	N_p	40
Probability of discovery	P_a	0.25
Upper and lower bounds of solution	$[l_b, u_b]$	±50
Maximum iteration	t_{max}	500
Maximum number of FE (function evaluation)	FE	500

5 Results

The optimization run has been performed in MATLAB 2015b. Since CSO is a stochastic optimization, a number of optimization runs need to be executed with different initial random seeds. To get an optimal solution and to evaluate the quality of the solution (robustness, convergence, repeatability), 20 runs have been executed. The statistical result of these runs is recorded in Table 4.

In general, a robust solution with a small standard deviation (good repeatability) can be achieved. The obtained feedback controller gains solution of those 20 runs is shown in Table 5. These controller gains obviously produce eigenvalues of the feedback system in a wedge region as specified in Fig. 3.

Furthermore, the mean value of the solution (controller gains) is taken from Table 5 and this will be called CSOFC (CSO-based feedback controller) for two-mass system. This CSOFC performance will be compared with that of DEFC (DE-based feedback controller) which has been designed and discussed previously in another study (Mahmud Iwan Solihin and Akmeliawati 2010). These two sets of controller gains are then listed in Table 6.

Table 4. Statistics of the 20 optimization runs

Average $f(X)$	−0.33
Median $f(X)$	−0.33
Standard deviation $f(X)$	0.004
Range of $f(X)$	−0.325 to −0.340
Average computation time	17 min (per run)

Table 5. Optimized feedback controller gains by CSO for two-mass system (20 runs)

Run no.	k_1	k_2	k_3	k_4	k_i
1	17.976	19.469	48.899	7.2406	−9.764
2	10.863	15.542	35.185	6.5217	−6.7774
3	14.899	19.403	45.902	7.4874	−8.7569
4	16.751	20.64	49.412	7.8372	−9.8067
5	17.992	19.923	49.23	7.4835	−10.125
6	17.091	20.236	48.985	7.6624	−9.8294
7	12.484	18.944	43.284	7.4513	−7.6838
8	15.774	20.592	48.921	7.8041	−9.2173
9	14.689	17.646	42.453	6.9183	−8.4257
10	17.429	19.66	48.416	7.4087	−9.7769
11	18.158	20.12	49.665	7.5478	−10.228
12	17.961	20.154	49.516	7.6	−10.107
13	17.008	20.43	49.77	7.6442	−9.6024
14	17.408	18.761	46.862	7.0931	−9.5514
15	15.782	20.609	49.255	7.7504	−9.1081
16	17.348	20.411	49.845	7.6449	−9.8158
17	17.21	19.347	47.868	7.281	−9.5449
18	15.007	18.499	44.452	7.1509	−8.6086
19	15.022	20.232	47.738	7.7165	−8.8319
20	16.566	19.611	47.253	7.5229	−9.6665
Mean	16.171	19.511	47.146	7.4383	−9.2614

Table 6. Feedback controller gains for two-mass system optimized using CSO and DE

Gains	k_1	k_2	k_3	k_4	k_i
CSOFC	16.17	19.51	47.15	7.44	−9.26
DEFC	18.49	19.04	47.27	7.30	−10.54

Figures 5 and 6 show 20 random samples of step response (position of the mass-2) for values of the spring constant $0.5 \leq k \leq 2$. The robustness performance of the proposed CSOFC under parameter variations of the plant is observed and comparison with that of DEFC is made. In addition, the visualized results show that CSOFC produces slightly more robust performance than DEFC. In term of the time-domain performance comparison for the feedback system with CSOFC and DEFC is shown in Table 7. It is clearly seen that CSOFC outperforms DEFC. i.e. the settling time (t_s) is just comparable and percentage of overshoot (PO) is smaller (both values are averaged from those 20 random step responses).

Another advantage of CS over DE, and perhaps over many other algorithms, is that only one parameter needs to be adjusted, i.e. probability of discovery (p_a). Meanwhile for DE, it needs at least two parameters to set, i.e. mutation factor and crossover constant.

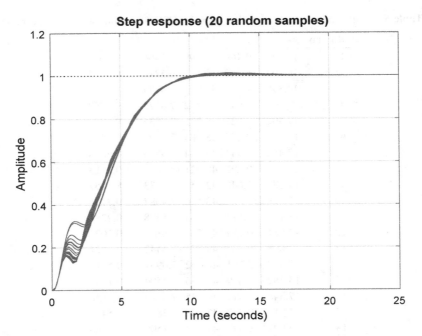

Fig. 5. 20 random step response of mass-2 displacement using CSOFC

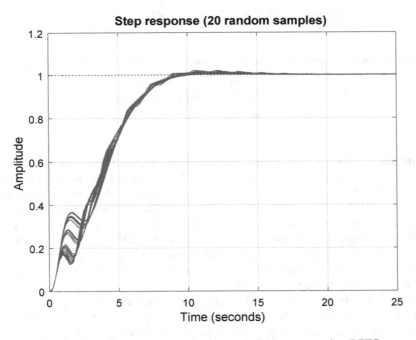

Fig. 6. 20 random step response of mass-2 displacement using DEFC

Table 7. Performance comparison for its step response (displacement of m_2)

Controller	t_s (sec)	PO (%)
CSOFC	9.1	1.1
DEFC	8.9	1.8

6 Conclusions

A robust state feedback control design via single objective constrained optimization using Cuckoo Search Optimization (CSO) to maximize stability radius has been proposed. The designed controller has shown the robust performance in the presence of parameter variations of the plant.

The results have also shown the effectiveness of CSO algorithm as compared to previously used DE algorithm to solve the same problem, the CSO-based feedback controller algorithm outperforms DE-based feedback controller in term of robustness performance. In Addition, another advantage of CSO over DE, and perhaps over many other algorithms, is that only one parameter needs to be adjusted in Cuckoo Search, i.e. discovery rate (p_a) in addition to other common parameters in metaheuristics algorithms such as number of population and number of iteration/generation.

References

Akmeliawati, R., Tan, C.P.: Feedback controller and observer design to maximize stability radius. In: 2005 IEEE International Conference on Industrial Technology, pp. 660–664. IEEE. http://ieeexplore.ieee.org/document/1600719/. Accessed 22 Aug 2018

Balochian, S., Ebrahimi, E.: Parameter optimization via Cuckoo optimization algorithm of fuzzy controller for liquid level control. J. Eng. **2013**, 1–7 (2013). http://www.hindawi.com/journals/je/2013/982354/ (August 20, 2018)

Barbosa, R.S., Jesus, I.S.: Optimization of control systems by Cuckoo search. In: Moreira, A.P., Matos, A., Veiga, G. (eds.) CONTROLO'2014. LNEE, vol. 321, pp. 113–122. Springer, Cham (2015). https://doi.org/10.1007/978-3-319-10380-8_12

Byrns, E.V., Calise, A.J.: Loop transfer recovery approach to H(Infinity) design for the coupled mass benchmark problem. J. Guidance Control Dyn. **15**(5), 1118–1124 (1992). http://arc.aiaa.org/doi/10.2514/3.20958. Accessed 21 Aug 2018

Coello Coello, C.A.: Theoretical and numerical constraint-handling techniques used with evolutionary algorithms: a survey of the state of the art. Comput. Methods Appl. Mech. Eng. **191**(11–12), 1245–1287 (2002). https://www.sciencedirect.com/science/article/pii/S0045782501003231. Accessed 22 Aug 2018

Hamza, M.F., Yap, H.J., Choudhury, I.A.: Cuckoo search algorithm based design of interval Type-2 Fuzzy PID controller for Furuta pendulum system. Eng. Appl. Artif. Intell. 62: 134–151 (2017). https://www.sciencedirect.com/science/article/abs/pii/S0952197617300714. Accessed 20 Aug 2018

Feyel, P.: Robust Control Optimization with Metaheuristics. ISTE (2017). https://www.wiley.com/en-us/Robust+Control+Optimization+with+Metaheuristics-p-9781786300423. Accessed 21 Aug 2018

Fister, I., Yang, X.-S., Fister, D., Fister, I.: Cuckoo search: a brief literature review. In: Yang, X. S. (ed.) Cuckoo Search and Firefly Algorithm. SCI, pp. 49–62. Springer, Cham (2014). https://doi.org/10.1007/978-3-319-02141-6_3

Gu, D.-W., Petkov, P., Konstantinov, M.M.: Robust Control Design with MATLAB. Springer, London (2005). https://doi.org/10.1007/978-1-4471-4682-7

Hinrichsen, D., Pritchard, A.J.: Stability radii of linear systems. Syst. Control Lett. **7**(1), 1–10 (1986). https://www.sciencedirect.com/science/article/pii/0167691186900940. Accessed 21 Aug 2018

Jin, Q., Qi, L., Jiang, B., Wang, Q.: Novel improved Cuckoo search for PID controller design. Trans. Inst. Meas. Control **37**(6), 721–731 (2015). http://journals.sagepub.com/doi/10.1177/0142331214544211. Accessed 20 Aug 2018

AI-Khafaji, A.A., Darus, I.Z.M.: Controller optimization using Cuckoo search algorithm of a flexible single-link manipulator. In: Proceedings of the 2014 First International Conference on Systems Informatics, Modelling and Simulation, pp. 39–44 (2014). https://dl.acm.org/citation.cfm?id=2681970.2682464. Accessed 21 Aug 2018

Kishnani, M., Pareek, S., Gupta, R.: Optimal tuning of PID controller by Cuckoo search via Lévy flights. In: 2014 International Conference on Advances in Engineering & Technology Research (ICAETR - 2014), pp. 1–5. IEEE (2014). http://ieeexplore.ieee.org/document/7012927/. Accessed 20 Aug 2018

Lu, H., Chen, W.: Dynamic-objective particle swarm optimization for constrained optimization problems. J. Comb. Optim. **12**(4): 409–419 (2006). http://link.springer.com/10.1007/s10878-006-9004-x. Accessed 22 Aug 2018

Reynoso Meza, G., Blasco Ferragud, X., Sanchis Saez, J., Herrero Durá, J.M.: The ACC'1990 control benchmark: a two-mass-spring system. In: Reynoso Meza, G., Blasco Ferragud, X., Sanchis Saez, J., Herrero Durá, J.M. (eds.) Controller Tuning with Evolutionary Multiobjective Optimization. ISCASE, vol. 85, pp. 147–157. Springer, Cham (2017). https://doi.org/10.1007/978-3-319-41301-3_7

Sethi, R., Panda, S., Sahoo, B.P.: Cuckoo search algorithm based optimal tuning of PID structured TCSC controller. In: Jain, L.C., Behera, H.S., Mandal, J.K., Mohapatra, D.P. (eds.) Computational Intelligence in Data Mining - Volume 1. SIST, vol. 31, pp. 251–263. Springer, New Delhi (2015). https://doi.org/10.1007/978-81-322-2205-7_24

Singh, K.S.M. Elamvazuthi, J.I., Shaari, K.Z.K., Lima, F.V.: PID tuning control strategy using Cuckoo search algorithm for pressure plant. In: 2016 6th International Conference on Intelligent and Advanced Systems (ICIAS), pp. 1–6. IEEE (2016). http://ieeexplore.ieee.org/document/7824127/. Accessed 20 Aug 2018

Solihin, M.I., Akmeliawati, R., Tijani, I.B., Legowo, A.: Robust state feedback control design via PSO-based constrained optimization. Control Intell. Syst. **39**(3), 168 (2011)

Solihin, M.I., Wen, M.C., Heltha, F., Lye, L.C.: Robust PID controller tuning for 2D gantry crane using Kharitonov's theorem and differential evolution optimizer. In: Advanced Materials Research, vol. 903 (2014)

Solihin, M.I., Akmeliawati, R.: Robust control design based on differential evolution for two-mass system. In: A Problem Statement. 2010, January 2015

Tang, S.H., Ang, C.K., Ariffin, M.K.A.B.M., Mashohor, S.B.: Predicting the motion of a robot manipulator with unknown trajectories based on an artificial neural network. Int. J. Adv. Rob. Syst. **11**(10), 176 (2014). http://journals.sagepub.com/doi/10.5772/59278. Accessed 21 Aug 2018

Tijani, I.B., et al.: Robust H-infinity controller synthesis using multi-objectives differential evolution algorithm (MODE) for two-mass-spring system. In: 2011 4th International Conference on Modeling, Simulation and Applied Optimization, ICMSAO 2011 (2011)

Wang, J., Li, S., Song, J.: Cuckoo search algorithm based on repeat-cycle asymptotic self-learning and self-evolving disturbance for function optimization. Comput. Intell. Neurosci. **2015**, 1–12 (2015). http://www.hindawi.com/journals/cin/2015/374873/. Accessed 22 Aug 2018

Wang, T., Meskin, M., Grinberg, I.: Comparison between particle swarm optimization and Cuckoo search method for optimization in unbalanced active distribution system. In: 2017 IEEE International Conference on Smart Energy Grid Engineering (SEGE), pp. 14–19. IEEE (2017). http://ieeexplore.ieee.org/document/8052769/. Accessed 20 Aug 2018

Wie, B., Bernstein, D.S.: Benchmark problems for robust control design. J. Guidance Control Dyn. **15**(5), 1057–1059 (1992). http://arc.aiaa.org/doi/10.2514/3.20949. Accessed 21 Aug 2018

Yang, X.S., Deb, S.: Engineering optimisation by Cuckoo search. Int. J. Math. Model. Numer. Optim. **1**(4), 330 (2010). http://www.inderscience.com/link.php?id=35430. Accessed 20 Aug 2018

Zabihi Samani, M., Amini, F.: J. Vibroeng. **17**. (2015). JVE International Ltd. https://www.jvejournals.com/article/15792. Accessed 20 Aug 2018

Chaos-Based Reinforcement Learning When Introducing Refractoriness in Each Neuron

Katsuki Sato, Yuki Goto, and Katsunari Shibata[✉]

Oita University, 700 Dannoharu, Oita, Japan
shibata@oita-u.ac.jp

Abstract. Aiming for the emergence of "thinking", we have proposed new reinforcement learning using a chaotic neural network. Then we have set up a hypothesis that the internal chaotic dynamics would grow up into "thinking" through learning. In our previous works, strong recurrent connection weights generate internal chaotic dynamics. On the other hand, chaotic dynamics are often generated by introducing refractoriness in each neuron. Refractoriness is the property that a firing neuron becomes insensitive for a while and observed in biological neurons. In this paper, in the chaos-based reinforcement learning, refractoriness is introduced in each neuron. It is shown that the network can learn a simple goal-reaching task through our new chaos-based reinforcement learning. It can learn with smaller recurrent connection weights than the case without refractoriness. By introducing refractoriness, the agent behavior becomes more exploratory and Lyapunov exponent becomes larger with the same recurrent weight range.

Keywords: Reinforcement learning · Chaotic neural network · Goal reaching · Refractoriness

1 Introduction

Our group has proposed the end-to-end reinforcement learning approach in which the entire process from input sensors to output motors that consists of a neural network is trained through reinforcement learning, and then various functions emerge in it [1]. DeepMind group showed the successful learning result of TV games in this approach [2], and that has consolidated the effectiveness of the approach. The higher functions that we human have, for example, memory, prediction, logical thinking need time-series processing. Our group has used a recurrent neural network that can learn to deal with time-series data, and has shown the emergence of memory or prediction function through reinforcement learning [3,4]. However, what we can call logical thinking, which is a typical higher function has not emerged yet.

We can think one after another without any input from outside, and so logical thinking can be thought of as internal dynamics that transit among states

© Springer Nature Singapore Pte Ltd. 2019
J.-H. Kim et al. (Eds.): RiTA 2018, CCIS 1015, pp. 76–84, 2019.
https://doi.org/10.1007/978-981-13-7780-8_7

autonomously. Exploration, which is essential in reinforcement learning, is similar to thinking in terms of dynamics with autonomous state transitions. From this similarity between logical thinking and exploration, we have set up a hypothesis that exploration, which is generated as chaotic dynamics, grows up into logical thinking through learning. So, we have proposed new reinforcement learning using a chaotic neural network (ChNN). Here, an agent explores according to its internal chaotic dynamics without adding random noises from outside, and can learn a simple goal-reaching task or an obstacle avoidance task [5,6]. In our previous works, strong recurrent connection weights generate internal chaotic dynamics in a recurrent neural network.

On the other hand, it is often the case that chaotic dynamics are generated by introducing refractoriness in each neuron [8]. Refractoriness is the property that neurons that have fired do not fire for a while, and is also the property that biological neurons actually have. Chaotic itinerancy, which we think very important property for inspiration or discovery, can be observed when associative memory is implemented. It is also shown that there is a difference in degree of chaos between known and unknown patterns on an associative memory using a ChNN, and after an unknown pattern is learned, association to the pattern is formed as well as the other known patterns [7].

In this paper, we introduce refractoriness in each neuron, and apply our new reinforcement learning to the refractoriness-originated chaotic neural network (RChNN). We examine whether the RChNN can learn a simple goal-reaching task. We compare the learning results between the cases of introducing refractoriness and without refractoriness, and observe Lyapunov exponent for both cases varying the range of the recurrent connection weights.

2 Reinforcement Learning (RL) Using a Refractoriness-Originated Chaotic Neural Network (RChNN)

In RL, an agent learns actions autonomously to get more a reward and less punishment. To realize autonomous learning, exploration is necessary, and in general, an agent explores stochastically using external random noises from outside. However here, an agent explores according to its internal chaotic dynamics in its ChNN without adding external random noises. For the learning of motor-level continuous motion signals, actor-critic is used. The actor-net, which generates actions, is made up of a ChNN, and the critic-net, which generates state value, is made up of a non-chaotic layered NN. The chaotic neuron model with refractoriness used in the actor net is dynamic as

$$u_{j,t}^{a,\xi} = \left(1 - \frac{\Delta t}{\tau}\right) u_{j,t-\Delta t}^{a,\xi} + \frac{\Delta t}{\tau} \sum_{i=1}^{N^{in}} w_{j,i}^{a,h} \cdot o_{i,t}^{a,in} \tag{1}$$

$$u_{j,t}^{a,\eta} = \left(1 - \frac{\Delta t}{\tau}\right) u_{j,t-\Delta t}^{a,\eta} + \frac{\Delta t}{\tau} \sum_{i=1}^{N^h} w_{j,i}^{a,REC} \cdot o_{i,t-\Delta t}^{a,h} \tag{2}$$

$$u_{j,t}^{a,\varsigma} = \left(1 - \frac{\Delta t}{\kappa\tau}\right) u_{j,t-\Delta t}^{a,\varsigma} - \frac{\Delta t}{\kappa\tau} \cdot \alpha \cdot o_{j,t-\Delta t}^{a,h} \tag{3}$$

$$u_{j,t}^{a,h} = u_{j,t}^{a,\xi} + u_{j,t}^{a,\eta} + u_{j,t}^{a,\varsigma} \tag{4}$$

$$o_{j,t}^{a,h} = \frac{1}{1 + \exp(-g \cdot u_{j,t}^{a,h})}. \tag{5}$$

The model is originated from the Aihara model [8], but we use the expression where the time constants are explicitly written. Each of Eqs. (1), (2) and (3) show forward, recurrent, or refractoriness term that appears in Eq. (4) respectively. $u_{j,t}^{a,h}$ and $o_{j,t}^{a,h}$ are the internal state and the output of the j-th hidden neuron at time t. $o_{i,t}^{a,in}$ is the i-th input signal. a indicates the actor-net, $h(=1)$ and $in(=0)$ indicate the hidden and input layer respectively. $w_{j,i}^{a,h}$ is the connection weight from the i-th neuron in the input layer to the j-th neuron in the hidden layer, and $w_{j,i}^{REC}$ is the recurrent connection weight from the i-th to the j-th neuron in the hidden layer. All the weights are decided by uniform random numbers. Here, we use step size $\Delta t = 1$, time constant $\tau = 1.25$, scaling parameter for time constant $\kappa = 8$ referring to [8], α is the scaling parameter of refractoriness, and g is the gain of sigmoid function. Here we use $\alpha = 3$ and $g = 2$. The neuron model used in the output layer ($L = 2$) in the actor-net or each neuron in the critic-net is static as

$$u_{j,t}^{n,l} = \sum_{i=1}^{N^{(l-1)}} w_{j,i}^{n,l} \cdot o_{i,t}^{n,l-1} \tag{6}$$

$$o_{j,t}^{n,l} = \frac{1}{1 + \exp(-u_{j,t}^{n,l})} - 0.5. \tag{7}$$

$n = a$ or c, and a represents the actor-net and c represents critic-net respectively. TD-error $\hat{r}_t$ used for learning is computed as

$$\hat{r}_t = r_{t+\Delta t} + \gamma \cdot V_{t+\Delta t} - V_t \tag{8}$$

where $r_{t+\Delta t}$ is the reward given at time $t + \Delta t$, γ is a discount factor, and here 0.96 is used. $V_{t+\Delta t} = O_{t+\Delta t}^{c,L}$ is the critic output, and $L(=2)$ represents the output layer. The training signal T_{V_t} for the output in the critic-net at time t is computed as

$$T_{V_t} = r_{t+\Delta t} + \gamma \cdot V_{t+\Delta t}. \tag{9}$$

The critic NN is trained once by regular error backpropagation using T_{V_t}.

In the proposed method, there is no external random number added to the actor outputs. The weights $w_{j,i}^{a,l}$ ($l = L(2), h(1)$) in the RChNN are modified using the causality trace $c_{j,i,t}^{l}$ [5] and a learning rate η as

$$\Delta w_{j,i,t}^{a,l} = \eta \cdot \hat{r}_t \cdot c_{j,i,t}^{l} \tag{10}$$

where $\Delta w_{j,i,t}^{a,l}$ is the update of the weight $w_{j,i}^{a,l}$. The trace $c_{j,i,t}^{l}$ is put on each connection, and takes and maintains the input through the connection according to the change in its output $\Delta o_{j,t}^{l} = o_{j,t}^{l} - o_{j,t-1}^{l}$ as

$$c_{j,i,t}^{l} = (1 - |\Delta o_{j,t}^{a,l}|)c_{j,i,t-\Delta t}^{l} + \Delta o_{j,t}^{a,l} o_{i,t}^{a,l-1}. \tag{11}$$

Here, only the connection weight $w_{j,i}^{a,l}$ from inputs to hidden neurons or from hidden neurons to output neurons are trained, and the recurrent connection weight $w_{j,i}^{a,REC}$ is not trained.

3   Simulation

In this paper, in order to examine whether our new reinforcement learning works also in an RChNN, we set a simple goal-reaching task as shown in Fig. 1 (Table 1).

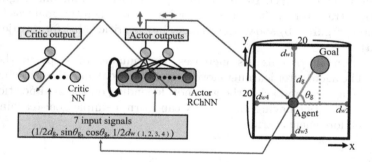

Fig. 1. Chaos-based reinforcement learning system in an agent and a goal-reaching task

In this simulation, as shown in Fig. 1, there is a 20×20 field. An agent with a radius of 0.5 and a goal with a radius of 1.0 are located randomly at the beginning of each episode. The agent catches 7 input signals representing the distance and direction from the agent to the goal and the distance to each wall, and inputs them to each neural network as in Fig. 1. Each of the two actor outputs represents the agent's movement in one step in the x-direction or the y-direction respectively, but they are normalized without changing its moving direction so that the movable range becomes a circle with a radius of 0.5. When the agent reaches the goal, it gains 0.4 reward. When the agent hits

Table 1. Parameters used in the simulation

		Actor	Critic
Number of hidden neurons		100	10
Gain of sigmoid function: g	Output	1	
	Hidden	2	1
Learning rate: η	Output	0.01	1
	Hidden (FW)	0.01	1
	Hidden (REC)	0	–
Range of initial weights	Hidden (FW)	$[-1, 1]$	
	Hidden (REC)	varied	–
	Output	$[-1, 1]$	

the wall, the agent is given a penalty of -0.01. One episode finishes when the agent reaches the goal or reaches 1,000 steps that is the upper limit of the step. The agent learned 50,000 episodes in total.

Figure 2 shows the learning curve. In order to see the early stage of learning, the horizontal scale is expanded in (a), while in order to see the late stage of learning the vertical scale is expanded in (b). Both red and blue lines show the number of steps to reach the goal but the red one is plotted at each episode while the average value over each 100 episodes is plotted as the blue one. Figure 3(a) shows sample trajectories after learning for 8 patterns when the goal was located at the center. Figure 3(b) shows the changes in the critic value for 8 trajectories in Fig. 3(a).

As the number of episodes increases, the number of steps to the goal decreases. The agent after learning moves toward the goal, and the critic becomes higher as the agent approaches the goal not depending on the goal position. We can see that an agent having an RChNN can learn a simple goal-reaching task with new reinforcement learning.

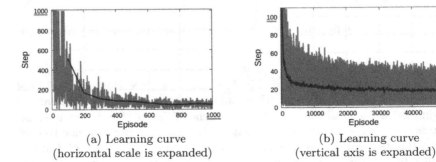

(a) Learning curve
(horizontal scale is expanded)

(b) Learning curve
(vertical axis is expanded)

Fig. 2. Learning curve (Color figure online)

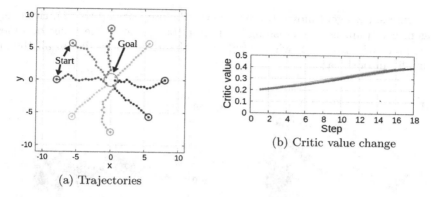

(a) Trajectories

(b) Critic value change

Fig. 3. 8 sample trajectories after learning and the critic value change for each trajectory

Next, we compare the learning success rate between an RChNN and a regular ChNN without refractoriness (the refractoriness term in Eq. (3) is removed). The range $[-w_{max}^{REC}, w_{max}^{REC}]$ of the recurrent connection weights w_{max}^{REC} was changed from 0.3 to 2, and the number of successful learning runs in 20 runs for each different weight range is compared as shown in Fig. 4. Regardless of having refractoriness, the number of unsuccessful runs increases as the recurrent connection weight decreases, but the success rate decreases faster in the case of without refractoriness.

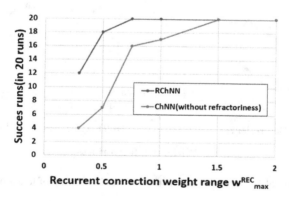

Fig. 4. Comparison of the learning success rate for the various range of recurrent connection weights in the hidden layer between the cases with and without refractoriness.

The agent behaviors before learning and after 100 episodes of learning for both cases when the weights range of the recurrent connection weights w_{max}^{REC} is 2.0 are shown in Fig. 5. We can see that the agent behavior is more exploratory with refractoriness than in the case without refractoriness. For the case without refractoriness, the agent moves toward the wall before learning (b−1). However, the agent gets punished by crashing, and after 100 episodes, the agent becomes more exploratory. In Fig. 4, when $w_{max}^{REC} = 2$, the agent succeeded in all the 20 runs finally in both cases.

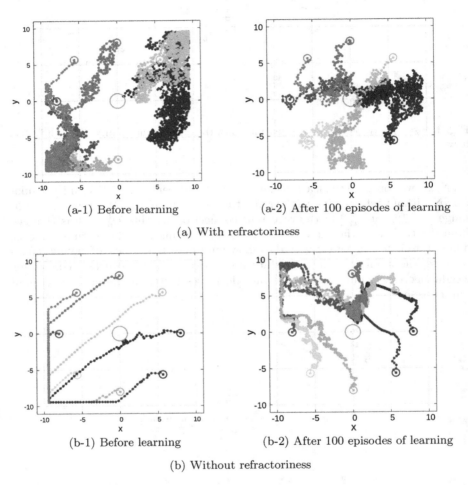

(a-1) Before learning   (a-2) After 100 episodes of learning

(a) With refractoriness

(b-1) Before learning   (b-2) After 100 episodes of learning

(b) Without refractoriness

Fig. 5. Comparison of the agent's exploration at the beginning of learning. ($w_{max}^{REC} = 2$)

In order to observe the relationship between degree of chaos and learning success rate, we calculate Lyapunov exponent. Lyapunov exponent is an index for degree of chaos in a dynamic system. If Lyapunov exponent is positive, the system is chaotic. Here, the network state is updated for 50 steps using only the recurrent connections, and is compared between the cases when a small perturbation whose size is 0.001 is added or not to initial state. Then Lyapunov exponent λ is calculated for 20 networks with different weights as

$$\lambda = \frac{1}{50 \cdot 20} \sum_{p=1}^{20} \sum_{t=1}^{50} \ln \frac{d_{p,t+\Delta t}}{d_{p,t}} \tag{12}$$

d is the distance in hidden state between the cases when a perturbation is added or not.

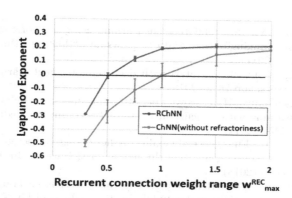

Fig. 6. Comparison of the Lyapunov exponent between the cases with or without refractoriness for the various ranges of the recurrent connection weights in the hidden layer.

In, Fig. 6, as the recurrent connection weights decreases, the degree of chaos decreases. The dynamics is more chaotic in the case with refractoriness than without refractoriness. The degree of chaos becomes stronger by introducing refractoriness. From the similarity of the trend between Figs. 4 and 6, introduction of refractoriness makes the degree of chaos strong and largely influences to the learning performance. We think that in the case with refractoriness, more exploratory behavior based on high degree of chaos is the source of the high successful learning rate for small recurrent connection weight ranges. Lyapunov exponent increases by using symmetrical activation function, but here we use Eq. (5).

4   Conclusion

In this paper, we used a refractoriness-originated chaotic neural network in our new reinforcement learning, and showed that the network can learn a simple goal-reaching task. As the recurrent connection weights decrease, the success rate decreases more slowly than the case without refractoriness, and that is very similar to the change in Lyapunov exponent. From the observation of the agent behavior at an early stage of learning, it is known that the introduction of refractoriness makes the degree of chaos strong and leads more exploratory behavior. That would be the reason why the success rate is larger in the case with refractoriness for the same range of the recurrent connection weights.

Acknowledgement. This work was supported by JSPS KAKENHI Grant Number 15K00360.

References

1. Shibata, K., Goto, Y.: New reinforcement learning using a chaotic neural network for emergence of "thinking" - "exploration" grows into "thinking" through learning. arXiv:1705.05551 (2017)
2. Volodymyr, M., et al.: Playing atari with deep reinforcement learning. In: NIPS Deep Learning Workshop 2013 (2013)
3. Shibata, K., Utsunomiya, H.: Discovery of pattern meaning from delayed rewards by reinforcement learning with a recurrent neural network. In: Proceedings of IJCNN 2011, pp. 1445–1452 (2011)
4. Shibata, K., Goto, K.: Emergence of flexible prediction-based discrete decision making and continuous motion generation through actor-Q-learning. In: Proceedings of ICDL-Epirob, ID 15 (2013)
5. Shibata, K., Sakashita, Y.: Reinforcement learning with internal-dynamics-based exploration using a chaotic neural network. In: Proceedings of IJCNN (2015)
6. Goto, Y., Shibata, K.: Emergence of higher exploration in reinforcement learning using a chaotic neural network. In: Proceedings of ICONIP 2016, pp. 40-48 (2016)
7. Osana, Y., Hagiwara, M.: Successive learning in chaotic neural network. In: Proceedings of IJCNN 1998, vol. 2, pp. 1510–1515 (1998)
8. Aihara, K., Takabe, T., Toyoda, M.: Chaotic neural networks. Phys. Lett. A **144**(6–7), 333–340 (1990)

Integration of Semantic and Episodic Memories for Task Intelligence

Ue-Hwan Kim and Jong-Hwan Kim[✉]

KAIST, 291 Daehak-ro, Yuseong-gu, Daejeon 34141, Republic of Korea
{uhkim, johkim}@rit.kaist.ac.kr
http://rit.kaist.ac.kr

Abstract. For successful completion of a task, an intelligent agent needs a carefully designed memory system which could enhance the autonomy of the whole system. Semantic and episodic memories play a key role in majority of memory systems by providing conceptual and sequential information. We propose a novel architecture that integrates both semantic and episodic memories for task intelligence. The adaptive resonance theory (ART) based episodic memory used in our work allows unsupervised learning of new experiences. The semantic memory developed in our architecture incorporates situational information as context. Considering the context enables the agent to generalize the situations where the learned-task can be performed. Thus, the integrated memory architecture progresses the autonomy and the performance of the agent improves even in new environments. To verify the performance of the proposed memory architecture, experiments are conducted under various conditions and the experiment results are analyzed.

1 Introduction

There are a number of key factors for an intelligent agent to successfully complete a given task. The key factors include mechanics [1–3], perception [4–6], planning [7–9], and memory [10, 11] systems. The intellectual components among the key factors for a task completion compose task intelligence [12, 13, 31, 32]. Among various components of task intelligence, the memory system plays a particularly crucial role. The memory system tells the agent what to do and how to do it and offers grounds for further inference and reasoning. Thus, careful design of the memory system is of great importance.

The memory system should be designed in a fashion that enhances the autonomy of the whole system. The autonomy, in the context of task intelligence, means that an agent can perform once-learned tasks in new environments that it never experienced. Without a structured memory system, the need for additional modules to enhance the autonomy arises since data from separate memory units need to be synthesized and processed for action planning [14, 15]. This increases system complexity while reducing system efficiency.

A variety of researches have contributed to the development of the field. The researches can be categorized in two groups. The first group focuses on representing tasks or environments in a generalized form and uses the representations to plan and

© Springer Nature Singapore Pte Ltd. 2019
J.-H. Kim et al. (Eds.): RiTA 2018, CCIS 1015, pp. 85–100, 2019.
https://doi.org/10.1007/978-981-13-7780-8_8

perform tasks in unseen environments [16–18]. However, the current level of symbolic representation and generalization of tasks and environments is immature to deal with sophisticated real-world situations. Another type of approach integrates semantic memory in planning [19–22]. This approach infers symbolic meaning and generalized knowledge from semantic memory. Nonetheless, it does not provide the fundamental solution as the approach detours around the same problem embedded in the first category. This approach just infers symbolic meanings from semantic memory. Furthermore, the second approach complicates the whole system since another unit is required to process the information from different parts.

In this paper, we propose a new memory architecture that efficiently stores task plans and relevant objects by integrating semantic and episodic memories. In the proposed memory architecture, the semantic memory encodes situational information as well as object information, which forms a context-based semantic memory architecture. The episodic memory saves the temporal sequence of a task plan whose components are encoded by the semantic memory. Equipping episodic memory with context-based semantic memory makes an intelligent system robust to changes in the environment, thus enhances the autonomy of the whole system. In addition, the integration scheme does not increase the system complexity because the unit itself is robust to differences between the learning and the actual environments and thus does not require an additional unit.

In summary, the contributions of our work are as follows.

1. We propose a new memory architecture that integrates both semantic and episodic memories in a manner that enhances the autonomy of the agent.
2. We formulate the concept of context to represent the situational information.
3. We develop encoding and decoding methodologies, which make the proposed memory system applicable for task intelligence and other intelligent systems.
4. We implement the memory system and verify the performance by conducting experiments under various situations.

The rest of the paper is organized in the following manner. Section 2 explains the concept of semantic and episodic memories used in the work. Section 3 presents the proposed memory architecture and Sect. 4 displays the experiment settings and the results with the analysis. Section 5 concludes the paper by presenting a few discussion points and concluding remarks.

2   Memory Structures

There exist multiple ways of classifying memory. Among them, we adopt the approach of Tulving [26] where semantic and episodic memories form the whole memory. Semantic and episodic memories differ in the types of information they store and the nature how they behave and function. In the following sub-sections, we describe each component of memory and the implementation scheme we take in construction of the integrated memory architecture.

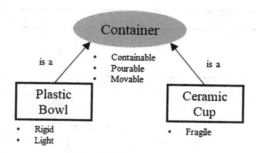

Fig. 1. An example of an ontology. An ontology consists of three representative primitives: classes, attributes, and relationships. Classes can have a hierarchical structure. 'Is a' describes the hierarchical relation among the classes. Here, container is the super-class of plastic bowl and ceramic cup and the dotted items represent the attributes. A sub-class inherits the attributes of the super-class.

2.1 Semantic Memory

Semantic memory is a highly structured network that stores concepts, words, and relations among concepts. The semantic memory does not register perceptible properties directly, but saves them after it processes the properties through generalization and abstraction. The memory system is independent of other memory systems once the system is formed and retrieval of information from the system does not change the content inside.

In the development of the proposed memory architecture, we implement the semantic memory as the form of an ontology. An ontology deals with a set of representative primitives to express the objects in the world. We use three types of representative primitives: classes, attributes, and relationships. Classes describe a name or the category of an object. Attributes represents properties of an object and relationships portrait the connectivity among class members. Figure 1 exemplifies an ontology. 'Is a' represents the relation between a super-class and a sub-class and both plastic bowl and ceramic cup are sub-classes of the container class. Each class has certain attributes and a sub-class inherits the attributes of the super-class. In the figure, the container class has three attributes and each sub-class of the container class inherits these attributes. Thus, plastic bowl has five attributes: containable, pourable, movable, rigid, and light. The ontology can be extended to multiple objects and include relations among the objects.

2.2 Episodic Memory

Episodic memory stores personal episodes or events and the temporal-spatial relations among the events. A temporal sequence of events forms an episode. In the process of receiving a new event, the memory encodes the event in terms of perceptible properties which are temporal-spatial relations and forms an autobiographical reference to the

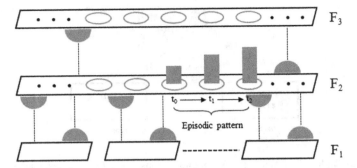

Fig. 2. The structure of an episodic memory based on the adaptive resonance theory (ART) network. The memory system contains three layers: input fields, event encoding layer, and episode encoding layer. The input fields receive binary or analog vectors and the input vectors make an element in the second layer fire. A series of events forms a pattern in the second layer and the pattern in the second layer gets encoded in the third layer as an episode.

already existing contents of the memory system. The episodic memory system is susceptible in the sense that retrieval of information from the memory changes the memory strength of the content inside. As one specific memory gets retrieved frequently, the memory becomes stronger. In the opposite case where a memory element is not retrieved for a long time, that memory gets forgotten.

We implement the episodic memory based on the adaptive resonance theory (ART) network [27]. ART was first introduced to explain the human cognitive information processing and later adopted for unsupervised learning and pattern recognition. The ART network can be extended to store and retrieve episodes [28] and we use a similar approach for the design of the episodic memory. Figure 2 depicts the structure of the episodic memory based on the ART network. The memory system has three layers and the layers are connected by appropriate weight vectors. The weights vectors are modified during the process of learning. The first layer contains multiple input fields. The input fields receive either binary or analog input vectors. The input vectors pass through the first layer and reach the second layer. In the second layer, only the element receiving the maximal scalar product of its weight vector and the input vector fires. The process so far encodes an event. The firing element decays as time goes on. After another event gets encoded, two elements of the second layer are activated as a result. The latter one has a higher value as the first one has gone through the decay process. This process continues until all the events of an episodes fire. Then, the pattern in the second layer which illustrates a series of events gets encoded in the third layer. The retrieval of a memory traverse the three layers in the opposite direction.

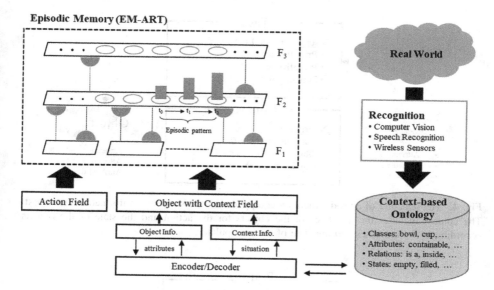

Fig. 3. The overall structure of the proposed memory architecture. The architecture integrates both semantic and episodic memories. The semantic memory is implemented as an ontology and the episodic memory unit is structured based on ART. The encoder/decoder unit generates the input vectors using the data coming from the context-based ontology. The decoder unit processes retrieved data from the episodic memory part and produces an output.

3 The Integrated Memory Architecture

3.1 The Integrated Architecture

Figure 3 shows the overall structure of the proposed memory architecture. The recognition unit perceives the real world and extracts symbolic meanings from the perceived data. The processed data go through the context-based ontology unit. The ontology unit supplements the data with additional information such as classes, attributes, relations, and states. After the ontology unit, the data packet contains context information. The encoder/decoder part translates the data packet from the ontology into input vectors and the ART-based memory structure receives the input vectors. The ART-based memory structure goes through learning processes. As described in Sect. 2, multiple events get learned in a temporal sequence to be stored as an episode. When a memory element is retrieved from the memory system, the data flows in the opposite direction and the processed data can be retrieved from the decoder unit.

3.2 Context-Based Semantic Mapping

In addition to attributes of an object, the situational information counts in task planning since the use of the object depends on the situation. The situational information, here, refers to the states of the objects and the environment. To give an example, an empty water bottle evokes the action of filling it or washing it while a filled water bottle evokes

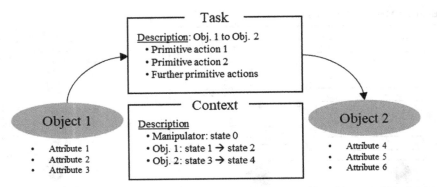

Fig. 4. The structure of episodic memory equipped with the situational information as context. The context includes the attributes of the objects for the action and the state of the objects according to the attributes and the state of the manipulator.

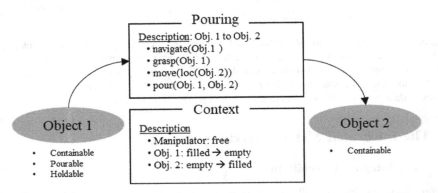

Fig. 5. An example of a memory element stored in the integrated memory architecture. The memory element memorizes objects as a quantity of attributes rather than a physical entity. This abstraction enhances the autonomy of an intelligent agent.

the action of pouring water to a cup or putting the water bottle into a refrigerator. As such, a different set of actions are meaningful for an object depending on different situations.

We define situational information as context. Figure 4 describes the concept of context. The context includes the states of the objects present in the world and that of the manipulator performing the task. The category of the object states are determined by its attributes. For instance, the container attribute has two states: being empty or filled. Another variable can be added to describe the situation in more detail. If the container is filled, for example, another state which illustrates the type of the volume it contains could be added.

Defining context helps an intelligent agent to symbolize the situation where the task can be performed. With the context, the agent views the objects as a quantity of attributes rather than a physical entity and the agents can perform the task even in new environments. Though not implemented in this work, the context can include the information regarding the shift of states. This extension allows an easy implementation of an efficient task planning algorithm.

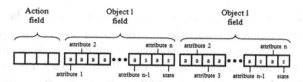

Fig. 6. The structure of an input vector for the proposed memory architecture. The input vector consists of three fields: action, object 1 and object 2. Each field represents the type of action and object with binary bits, respectively. The object field contains the context information.

3.3 Encoding and Decoding Scheme

An event equipped with the context gets encoded in the form of binary vector. The input vector consists of three fields. Each three fields contain the information regarding action, object 1, and object 2, respectively. The context information is already contained in object fields. We adopt one-hot vector representation for the encoding. Therefore, the number of bits of each field is determined by the variety of situations the system needs to deal with. For example, the number of actions available for an agent determines the number of bits in the action field. If the agent could perform sixteen types of actions, sixteen bits are required for the action field. The object fields describe the attributes of the objects rather than the name of the objects used for the action. The attributes specify the necessary conditions for the action to be performed. Thus, any objects possessing the required attributes can be used for the action. Representing objects as a quantity of attributes rather than the name symbolizes the concept of the action. In this work, each object field contains eight attribute sections and three of the attribute sections have a tail bit to represent the state of the attribute. The tail bit for the attribute 'containable', for instance, illustrates the object is empty when the bit is zero and the object is filled when the bit is one. The structure of an input vector is described in Fig. 6. Each square represents a bit.

3.4 Learning the Sequence of a Task Episode

The temporal sequence of input vectors encoded with the context information enter the episodic memory part. The episodic memory part learns how to perform a task episode. As the proposed episodic memory is based on the ART network, it learns a new task in an incremental manner. In other words, the memory architecture can learn new sequences without forgetting and re-learning previously learned task episodes. The episodic memory consists of two ART networks. The first layer classifies the input events and the sequence of events forms an episodic pattern in F_2 layer. The second layer learns the episodic pattern in F_2 layer and stores them in F_3 layer.

The ART network compares the input vector with the stored memories. The code activation process defines the comparison process:

$$T_j = \sum_{k=1}^{n} \gamma^k \frac{\left| x^k \bigwedge w_j^k \right|}{\alpha^k + |x^k|},\tag{1}$$

where j is the node index, k is the input channel index, w_j^k is the weight of the network, x^k is the activity vector, $\bigwedge$ is a fuzzy-min operator, and α is a choice parameter. The node with the maximum activation is chosen to check the resonance condition. When the condition is met, the parameters are updated. When the new input vector does not satisfy the resonance condition, a new category is created in the memory and the weights of the new node are set as the input vector. In this work, we use the approach in Deep ART [34] to generate and encode event patterns in the F_2 layer, which get stored in the F_3 layer as a task episode. For the further description of ART network and the pattern generation methods, one can refer to [28].

After the network learns a new task, the task is stored as a memory element. Figure 5 depicts an example of a memory element stored in the proposed memory architecture. The memory element was learned from the task episode which describes pouring water into a cup from a water bottle. Although the episode used for learning specifies object 1, 2 and the material inside object 1, the memory element illustrates the generalized situation where the task of pouring can be performed. In other words, the agent can perform the task in general situations once a task is learned in one environment. In addition, two objects are needed to perform the pouring task: one object that is 'containable', 'pourable', and 'movable' and another object that is 'containable'. The first object should have the state of 'filled', otherwise pouring does not have any meaning.

3.5   Application in Task Intelligence

The integrated memory architecture applies to the task intelligent settings, where various types of tasks are performed by an intelligent agent. The types of tasks include 'watering a flower pot', 'making cereal', 'serving coffee', 'finding an object from a drawer', and 'sorting toys'. The agent learns how to complete these tasks in one environment and needs to achieve similar performance in general situations. If the agent could not perform in other environments but do only in the same environments, the learning becomes meaningless. The proposed memory architecture, which enhances the autonomy of an intelligent agent, can be employed to guarantee the best performance of the agent in task intelligence settings.

4   Experiments

4.1   Experiment Setup

To verify the performance of the proposed memory architecture, we conducted experiments under various conditions. The experiments were composed of two phases: the learning phase and the retrieval phase. In the learning phase, we varied the learning condition of the proposed memory architecture by controlling the number of training

Table 1. The list of actions used in the experiments.

Type	Name	Parameter 1	Attribute 1	Param. relation	Parameter 2	Attribute 2
elementary action	move	(loo) A	null			
	navigate	(name) A	null			
	grasp	(obi) A	holdable			
	open	(obj) A	openable			
complex action	put	(obj) A	holdable, puttable	into	(obj) A	containable
	put down	(obj) A	holdable	on	(obj) A	standable
	find	(name) A	null	in	(obj) A	containable
	pour	(obj) A	holdable, pourable	into	(obj) A	containable
	water	(obj) A	waterable	with	(obj) A	spray

Table 2. The list of attributes used in the experiments

Name	State
Null	–
holdable	–
puttable	–
pourable	–
waterable	–
spray	–
containable	empty/tilled
standable	free/full
openable	open/closed

epochs. We changed the number of training epochs to analyze the effect of the training frequency on the retrieval accuracy. We trained the memory system with episodes from task intelligence settings such as 'pouring water into a cup from a water bottle'.

The encoder transforms each event in a task episode into an input vector using the information from the context-based ontology. The categories of actions and the attributes with the possible states are listed in Tables 1 and 2. In the experiments, we used nine actions which can be classified into two types. Four elementary actions take one parameter and five complex actions take two parameters. The parameter relation represents the relation between two parameters in complex actions. In addition, we used nine types of attributes for the experiments. Six of the attributes do not accompany with state information and the rest require one bit for describing state information. We allocated one bit to represent the state of the manipulator.

In the retrieval phase, we tested the proposed memory architecture in two situations. Firstly, we assessed if the proposed memory system could retrieve the proper event sequences for completing the given task in the same environment where the task had been learned. This step checked the basic functionality of a memory system: retrieving learned tasks. Furthermore, we evaluated if the proposed memory architecture could perform the given task in new environments. We substituted the objects with other objects that had different properties than the original objects and tested if the memory architecture could retrieve the relevant episode.

For the comparative study of the proposed method, we used two approaches: an episodic memory structure without semantic memory (the baseline ART-based memory) and long short-term memory (LSTM) networks [33]. The episodic memory structure without semantic memory on which our work is based has a similar structure as our proposed memory architecture. However, the baseline structure does not process and encode context information. In our experiments, we controlled two parameters of LSTM: the number of layers and the size of hidden layer. The size of hidden layer as well as the number of layers determine the degree of generalization. Although the number of training epochs affects the retrieval accuracy for LSTM, we only evaluated the loss over training epochs. Both memory structures for the comparative study shared the same recognition unit and the data processing module with the proposed method.

When training LSTM networks, the networks do not explicitly save an memory element but the memories are stored in the connections among the nodes. Thus, one cannot retrieve a specific sequence from the networks. When testing with LSTM networks, we gave a partial cue to the networks and assessed if the networks could retrieve the rest of the relevant sequence. The same test method can be applied to the ART based memories. Both the proposed memory architecture and the baseline method can accept a partial cue and retrieve the relevant memory element. We also evaluated the proposed method with the same partial cue approach.

4.2 Results and Analysis

The encoder generated the same input vector for the same type of events in the proposed memory architecture. For example, the event of 'pouring water into a cup from a water bottle', the event of 'pouring juice into a glass from a jar', and the event of 'pouring milk into a bowl from a milk bottle' all resulted in the same input vector

Table 3. The experiment results

Experiment	Method					
	Proposed method	EM-ART	LSTM L1H20	LSTM L1H40	LSTM L2H20	LSTM L2H40
Task in the learning environment	47/47	47/47	–			–
Task in a new environment	47/47	0/47	–			–
One-by-One		–	38/42	38/42	37/42	37/42
Partial cue (50%)	20/23	–	11/15	11/15	10/15	10/15
Partial cue (30%)	20/30	–	19/23	19/23	18/23	18/23

because the objects have the same attributes, which is the outcome of the enhanced autonomy. Furthermore, the same type of actions were generalized and classified as the same event. After all, the encoder produced nine types of input vectors for each available action. Then, we trained the proposed memory architecture with five tasks. Each task episode length varied from five to sixteen. Figure 7 shows the training procedure. The horizontal axis represents the number of events the memory network learned. The vertical axis shows the number of actions (red dots) and the scenario (blue dots). The summation of events in all scenarios was forty-seven. As mentioned above, the proposed memory architecture learned events and scenarios incrementally. After one iteration of training, the proposed memory system successfully learned all the actions and the scenarios. Furthermore, the proposed system can classify all the actions after one training epoch. We modified the number of training epochs from one to ten, but they all showed the same performance. One iteration sufficed because the numbers of actions and the scenarios were small. If the numbers had been larger, the training might have taken longer.

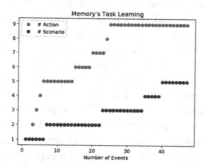

Fig. 7. The training procedure of the proposed memory architecture. As the number of events the system learns increases, the total number of actions and scenarios in the architecture increment. (Color figure online)

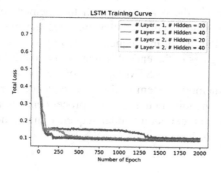

Fig. 8. The training curves of the LSTM variants used in the experiments. After 500 training epochs, the LSTM variants except L1H20 converged. L1H20 converged after 1,300 training epochs.

For the design of LSTM, the number of hidden layers and the size of hidden layer were varied. In our experiments, LSTM possessed one or two hidden layers with the size of 20 and 40. They are abbreviated as LSTM L1H20, L1H40, L2H20 and L2H40. For the training of LSTM variants, same encoder used for the proposed memory architecture was employed. Therefore, LSTM variants and the proposed memory system shared the same input vectors. Figure 8 depicts the training curve for the four LSTM variants. Except L1H20, LSTM variants converged after 500 training epochs. However, L1H20 took around 1,300 training epochs for convergence. The generalization process took longer with smaller hidden layer.

Table 3 summarizes the experiments results. In the table, the retrieval accuracies are represented in event units rather than task units. We conducted the first two experiments to compare the proposed memory architecture with the baseline ART-based memory. Both memory systems were trained with the same tasks but only the first system processed input data through the encoder. When the memory systems were tested in the learning environment, both could retrieve the relevant task episodes. However, changes in the environment hindered the ART-based memory from retrieving the proper memory element. The baseline memory architecture does not store generalized memory elements and only provides specific cases of experienced memories due to lack of context-awareness property. In contrast, the proposed memory architecture successfully generalizes tasks using context-based ontology.

For the one-by-one experiment, elements of tasks were given to each LSTM variants one by one and the output sequences were compared with the actual task sequences. For example, assume task 1 consists of event 1 through event 8. For the evaluation with task 1, event 1 through event 7 were given to the LSTM and the output sequence was compared with the sequence composed of event 2 through event 8. LSTM variants showed similar performances and were not able to retrieve the same part of the test sequences. Three of tasks used in the experiment have the same starting sub-sequence and the LSTM variants could not distinguish the three tasks until the different part appear.

The last two experiments were to evaluate if the memory systems could retrieve the full sequence out of partial cues. We used two cue lengths. When the half of the sequences were given to the memories, the proposed method performed the best. The proposed method compares the partial cue with the memory elements stored in the system and extracts the whole sequence that matches the best. On the other hand, LSTM needs to extract the rest of the sequences step by step and one error during the extraction process propagates to the end. When around one third of the sequences were given to the systems as a cue, the LSTM variants outperformed the proposed memory system. The proposed memory system retrieves a sequence in a memory unit, i.e. an episode unit. Therefore, one miss in the retrieval process ends in the error for the whole task. LSTM, in contrast, can extract a similar sequence that is not correctly match the true sequence.

5 Discussion

In this section, we present a few discussion points. Firstly, the proposed memory architecture enables an intelligent agent to learn how to perform a task in an unsupervised way. The proposed memory architecture classifies the types of tasks and groups the same type of tasks in a cluster. In addition, the ART structure on which our proposed method is based enables the memory structure to learn tasks in a incremental manner. If an instructor teaches the intelligent agent from time to time, the agent with the proposed memory architecture could learn new tasks without forgetting previously learned tasks. In summary, the agent could learn an unlimited number of tasks and sort them accordingly once the recognition part of the agent functions correctly and the ontology offers all the relevant information.

Furthermore, an intelligent agent with the proposed method learns both how to perform a task and the generalized situation where the task can be performed. Unlike sophisticated generalization methods, the proposed method uses an ontology to symbolize the situation where a task can be performed. Object attributes, states and manipulator states extracted from the ontology form the context. This context lets the agent to determine which objects and situations are relevant for performing a task. This aspect together with the above mentioned property enhances the autonomy of the agent.

Next, the integrated memory architecture can be combined with various types of ART-based memory systems due to the flexibility and modularity of the architecture. Other ART-based memory systems can support the learning procedure and provide semantic information. The proposed memory architecture can incorporate those ART-based memories by connecting them to the system and slightly modifying the input module. The input module consists of encoder, decoder and context-based ontology in addition to the conventional input field of action.

Although the proposed method showed satisfactory performance, there still exists a room for further improvement. First of all, the memory architecture should be able to generalize actions as the situations and the objects were generalized. The proposed method describes situations and objects with attributes and situational information which form the context. Actions, in a similar way, can be generalized as a quantity that exerts an effect to an object. The concept of affordance formalizes this idea [29, 30, 35]. Affordance represents an action with the effects the action exerts on an object. By adopting the concept of affordance, the memory architecture will become even more flexible and robust.

We discuss the next point of further improvements. In this study, we verified the effectiveness and applicability of the proposed method by presenting the abstraction power of the memory system using computer simulations. The proposed memory architecture should be applied to real-world scenarios where physical intelligent agents perform a temporal sequence of primitive actions to complete a task. For this, we need to embed the memory architecture in real agents such as robots and demonstrate that the proposed memory architecture enhances the autonomy of the agents in general environments. In the following research, a thorough experiment with a robot would be done to further verify the proposed method.

6 Conclusion

In this paper, we proposed a new memory architecture that integrates both semantic and episodic memories for the enhancement of autonomy of an intelligent agent. The semantic memory incorporates situational information with ontology primitives which defines the context. The context information is attached to each event of an episode. Equipping episodic memory with context-based semantic memory enables an intelligent agent to conceptualize the situation where a task can be executed. These processes improve the autonomy of the agent. We showed the performance of the proposed method through thorough experiments and discussed future research directions.

Acknowledgement. This work was partly supported by Institute for Information & communications Technology Promotion (IITP) grant funded by the Korea government (MSIT) (No. 2018-0-00677, Development of Robot Hand Manipulation Intelligence to Learn Methods and Procedures for Handling Various Objects with Tactile Robot Hands) and by the National Research Foundation of Korea (NRF) grant funded by the Korea government (MSIT) (No. NRF-2017R1A2A 1A17069837).

References

1. Baldonado, M., Chang, C.-C.K., Gravano, L., Paepcke, A.: The Stanford digital library metadata architecture. Int. J. Digit. Libr. **1**, 108–121 (1997)
2. Bruce, K.B., Cardelli, L., Pierce, B.C.: Comparing object encodings. In: Abadi, M., Ito, T. (eds.) TACS 1997. LNCS, vol. 1281, pp. 415–438. Springer, Heidelberg (1997). https://doi.org/10.1007/BFb0014561
3. van Leeuwen, J. (ed.): Computer Science Today: Recent Trends and Developments. LNCS, vol. 1000. Springer, Heidelberg (1995). https://doi.org/10.1007/BFb0015232
4. Michalewicz, Z.: Genetic Algorithms + Data Structures = Evolution Programs, 3rd edn. Springer, Heidelberg (1996). https://doi.org/10.1007/978-3-662-03315-9
5. Craig, J.J.: Introduction to Robotics: Mechanics and Control, 3rd edn. Prentice Hall, Upper Saddle River (2003)
6. Tsai, L.-W.: Robot Analysis: The Mechanics of Serial and Parallel Manipulators. Wiley, New York (1999)
7. Channon, P.H., Hopkins, S.H., Pham, D.T.: Derivation of optimal walking motions for a bipedal walking robot. Robotica **10**, 165–172 (1992)
8. Elfes, A.: Using occupancy grids for mobile robot perception and navigation. Computer **22**, 46–57 (1989)
9. Correa, J., Soto, A.: Active visual perception for mobile robot localization. J. Intell. Robot. Syst. **58**, 339–354 (2009)
10. Friedman, L.: Robot perception learning. ACM SIGART Bull. (1977)
11. Brooks, R.A.: Symbolic error analysis and robot planning. Int. J. Robot. Res. **1**, 29–78 (1982)
12. Qureshi, A.H., Ayaz, Y., Osaka: Reference tools. Auton. Rob. 1–15 (2015)
13. McDermott, D., Davis, E.: Planning routes through uncertain territory. Artif. Intell. **22**, 107–156 (1984)
14. Schaal, S., Atkeson, C.G.: Robot juggling: implementation of memory-based learning. IEEE Control Syst. **14**, 57–71 (1994)

15. Menegatti, E., Maeda, T., Ishiguro, H.: Image-based memory for robot navigation using properties of omnidirectional images. Rob. Auton. Syst. **47**, 251–267 (2004)
16. Park, G.-M., Kim, J.-H.: Deep adaptive resonance theory for learning biologically inspired episodic memory. In: 2016 International Joint Conference on Neural Networks (IJCNN), Vancouber (2016)
17. Beetz, M., et al.: Cognition-enabled autonomous robot control for the realization of home chore task intelligence. Proc. IEEE **100**(8), 2454–2471 (2012)
18. Kim, J.-H., Choi, S.-H., Park, I.-W., Zaheer, S.A.: Intelligence technology for robots that think [application notes]. IEEE Comput. Intell. Mag. **8**, 70–84 (2013)
19. Yoo, Y.-H., Kim, J.-H.: Procedural memory learning from demonstration for task performance. In: 2015 IEEE International Conference on Systems, Man, and Cybernetics (SMC), Kowloon, pp. 2435–2440 (2015)
20. Park, J.-Y., Yoo, Y.-H., Kim, D.-H., Kim, J.-H.: Integrated adaptive resonance theory neural model for episodic memory with task memory for task performance of robots. In: 2016 International Joint Conference on Neural Networks (IJCNN), Vancouber (2016)
21. Ghallab, M.: PDDL - The planning domain definition language V. 2. Technical report, report CVC TR-98–003/DCS TR-1165, Yale Center for Computational Vision and Control (1998)
22. Fox, M., Long, D.: PDDL2.1: an extension to PDDL for expressing temporal planning domains. J. AI Res. **20**, 61–124 (2003)
23. Nieoiescu, M., Mataric, M.: Natural methods for robot task learning: instructive demonstrations, generalization and practice. In: Proceedings of International Joint Conference on Autonomous Agents and Multiagent Systems (AAMAS), pp. 241–248 (2003)
24. Galindo, C., Fernandez-Madrigal, J., Gonzalez, J.: Improving efficiency in mobile robot task planning through world abstraction. IEEE Trans. Rob. **20**, 677–690 (2004)
25. Harnad, S.: The symbol grounding problem. Physica D: Nonlinear Phenomena **42**, 335–346 (1990)
26. Galindo, C., Saffiotti, A., Coradeschi, S., Buschka, P., Fernández-Madrigal, J., Gonzalez, J.: Multi-hierarchical semantic maps for mobile robotics. In: Proceedings of IEEE/RSJ International Conference on Intelligent Robots and Systems, pp. 2278–2283 (2005)
27. Galindo, C., Fernandez-Madrigal, J., Gonzalez, J., Saffiotti, A.: Using semantic information for improving efficiency of robot task planning. In: ICRA Workshop: Semantic Information in Robotics (2007)
28. Beeson, P., MacMahon, M., Modayil, J., Murarka, A., Kuipers, B., Stankiewicz, B.: Integrating multiple representations of spatial knowledge for mapping, navigation, and communication. In: Interaction Challenges for Intelligent Assistants, pp. 1–9 (2007)
29. Galindo, C., Fernández-Madrigal, J., González, J., Saffiotti, A.: Robot task planning using semantic maps. Rob. Auton. Syst. (2008)
30. Tulving, E.: Episodic and semantic memory 1. In: Organization of Memory, vol. 381, no. 4. Academic, London (1972)
31. Carpenter, G., Grossberg, S., Rosen, D.: Fuzzy ART: an adaptive resonance algorithm for rapid, stable classification of analog patterns. In: Proceedings of 1991 International Joint Conference on Neural Networks (1991)
32. Wang, W., Subagdja, B., Tan, A.-H., Starzyk, J.A.: Neural modeling of episodic memory: encoding, retrieval, and forgetting. IEEE Trans. Neural Netw. Learn. Syst. **23**, 1574–1586 (2012)
33. Chemero, A.: An outline of a theory of affordances. Ecol. Psychol. **15**, 18–195 (2003)
34. Sahin, E., Cakmak, M., Dogar, M., Ugur, E., Ucoluk, G.: To afford or not to afford: a new formalization of affordances towards affordance-based robot control. Adapt. Behav. **15**, 447–472 (2007)

35. Jeong, I.-B., Ko, W.-R., Park, G.-M., Kim, D.-H., Yoo, Y.-H., Kim, J.-H.: Task intelligence of robots: neural model-based mechanism of thought and online motion planning. IEEE Trans. Emerg. Top. Comput. Intell. **1**, 41–50 (2017)
36. Kim, D.-H., Park, G.-M., Yoo, Y.-H., Ryu, S.-J., Jeong, I.-B., Kim, J.-H.: Realization of task intelligence for service robots in an unstructured environment. Annu. Rev. Control **44**, 9–18 (2017)
37. Hochreiter, S., Schmidhuber, J.: Long short-term memory. Neural Comput. **9**, 1735–1780 (1997)
38. Park, G.-M., Yoo, Y.-H., Kim, D.-H., Kim, J.-H.: Deep ART neural model for biologically inspired episodic memory and its application to task performance of robots. IEEE Trans. Cybern. **48**, 1786–1799 (2017)
39. Han, J.-H., Lee, S.-J., Kim, J.-H.: Behavior hierarchy-based affordance map for recognition of human intention and its application to human-robot interaction. IEEE Trans.-Hum. Mach. Syst. **48**, 1786–1799 (2016)

Human Robot Social Interaction Framework Based on Emotional Episodic Memory

Won-Hyong Lee, Sahng-Min Yoo, Jae-Woo Choi, Ue-Hwan Kim, and Jong-Hwan Kim[✉]

School of Electrical Engineering, KAIST (Korea Advanced Institute of Science and Technology), Daejeon 34141, Republic of Korea
{whlee, smyoo, jwchoi, uhkim, johkim}@rit.kaist.ac.kr

Abstract. Nowadays the application of robots are emerging in areas of modern life. It is expected we will be living in a new era in which robots such as socially interactive robots will make an important effect on our daily lives. Considering emotions play a critical role in human social communication, emotional episodes are necessary for human-robot social interactions. In this regard, we propose a framework that can form a social relationship between human and robot using an emotional episodic memory. Proposed framework enables personalized social interactions with each user by identifying the user and retrieving the matching episode in the memory. The interaction is not fixed, the emotional episodic memory is developmental through additional experiences with the user. The proposed framework is applied to an interactive humanoid robot platform, named Mybot to verify the effectiveness. As demonstration scenarios, photo shooting, and user identification and robot's emotional reactions are used.

1 Introduction

Recently, the development of Artificial Intelligence (AI), Internet of Things (IoT), and Cloud technologies has been driving the growth of the social robot market. Robots are getting closer to people and assist people's everyday life. It is expected that social robots will have a very significant impact in the near future [1, 2]. International Federation of Robotics (IFR) predicted social robots would be commercialized and provide actual services to people between 2015 and 2018. As well as technical factors, fast development of aging society and increase in single-person households demand for the social robots that can provide mental and emotional services. Moreover, people have a tendency of assigning socially relevant characteristics to any device [3], and people have social expectations and want to be socially involved with robots [4].

Robots with social interaction systems have been developed for years. Bartneck and Forlizzi defined social robot as an autonomous or semi-autonomous robot that interact and communicates with humans by following the behavioral norms expected by the people with whom the robot interacts [5]. Various frameworks for HRI have been proposed. Such frameworks include a framework employing multi-modal for utilizing various types of information [6, 7], a framework for working with programmer and interaction designer together [8], and a framework for a robust robot control [9].

© Springer Nature Singapore Pte Ltd. 2019
J.-H. Kim et al. (Eds.): RiTA 2018, CCIS 1015, pp. 101–116, 2019.
https://doi.org/10.1007/978-981-13-7780-8_9

Emotional interaction lies at the core of building social relationships [10]. People perceive a robot that can express emotions as more anthropomorphic [11] and can maintain a more intimate relationship with such robot. Several researches attempted to reflect emotions on robots. These researches mainly consider frameworks for generating emotions appropriate to the situation and expressing them effectively [12–15]. A study that the robot provides appropriate services according to the emotions of users was also reported [16].

However, previous studies lack an emotional memory module for human robot social interaction. This leads to two limitations. Firstly, a social interaction framework without an emotional memory module cannot differentiate users, and thus it cannot provide a personalized services. Secondly, robots with the previous frameworks rarely remember earlier interactions with the user. Therefore, such robots only provide the same interactions repeatedly and the relationship with the user cannot be further developed.

To overcome the above-mentioned limitations, we propose a framework for human robot social and emotional interaction based on emotional episodic memory. We use a hierarchical emotional episodic memory (HEEM) [17], where the robot can store interaction experiences with each user and the entailing emotions. A robot with the proposed framework can provide personalized services to different users with proper emotions. The proposed framework enables the robot to have different relationships with each person. Moreover, emotions generated from our robot by the framework are not fixed, but gradually develop through interaction experiences. Thus, the robot and users can establish a more natural social relationship. We implemented an interactive hardware platform, Mybot to verify the performance of the proposed framework.

The rest of the paper is organized as follows. Section 2 describes the details of proposed framework architecture. Section 3 represents designed scenarios to demonstrate the feasibility and reliability of the framework with robotic platform and conducted surveys from users. Sections 4 and 5 interpret the survey results from users and discuss further work. Specific software and hardware design details of our interactive robotic platform are described in APPENDIX.

2 Proposed Framework Architecture

Figure 1 shows the proposed framework architecture for the human robot social interaction. The architecture is composed of (1) sensory part, (2) recognition part, (3) language part, (4) memory part, (5) communication part, (6) control part, (7) cloud applications. The robot receives user's face, voice, touch as input, and makes social interaction possible through the appropriate reaction output. Sensory part collects input data using image, sound, and touch sensor.

Recognition part consists of modules that analyze the data transmitted from the sensory part and discerns what the data mean. It finds out who the user is (User face identification), figure out the user's expression (Face Expression Recognition), and apprehends user's speech (speech recognition) and touch action (touch recognition). Analyzed information is sent to language part and memory part for further processing.

Language part gets the text extracted from the user's dialogue and figures out the meaning of it to get the context and the status that the user wants to convey to the robot. After that, Dialogue generator module makes an appropriate response during autonomous conversation with the user. These factors are sent to the memory part to let the memory part knows what conversation the user had with the robot.

Memory part stores the emotional episodes with the user, generates robot's emotion so that the user and the robot can communicate emotionally. Working memory module makes the most appropriate decision among some choices to continuously maintain the social relationship.

Communication part and the control part express the reactions directly to the users whereas previous modules were responsible for deciding which reactions to provide to the user. Communication part plays the synthesized voice of the robot and shows the robot's facial expression so that user can naturally hear and see the reply of the robot. Control part regulates robot's neck movement and body motion.

In order to enable the above mechanism, several cloud applications are used. The implementation details in APPENDIX gives comprehensive descriptions of Cloud applications.

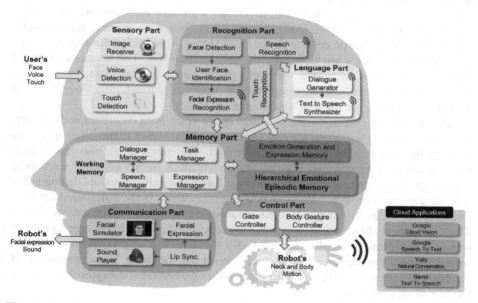

Fig. 1. The proposed human robot social interaction framework architecture. It consists of sensory part, recognition part, language part, memory part, communication part, control part, and cloud applications. Detailed description is given in Sect. 2

2.1 Sensory Part

HRI starts from perceiving the user and the environment. Sensory part collects data for understanding current state of the user by face, voice, and touch data. Furthermore, the sensory part receives the input signal from the users and transmits to the other modules to extract the meaning of the signal.

Sensory part consists of three modules: image receiver module, voice detection module, and touch detection module. The image receiver module obtains camera images of the point of view from the robot. The images are mainly used for identifying the user and user's facial expression. The voice detection module detects human voice. We design the voice detection module in a way that recording the voice when sound level exceeds a certain threshold for the energy efficiency. The collected sound data are transmitted to the speech recognition module to be analyzed. The touch detection module receives the user's touch data. In the proposed framework, touch data is used for patting or bullying the robot, which affects the robot's emotion formation.

2.2   Recognition Part

The recognition part processes the raw data from sensory part and extracts information needed for social interaction. The recognition part consists of three modules: face detection module, speech detection module, and touch detection module.

Face detection module recognizes two things, identifies who the current user is and what facial expression appears on the user's face. User identification among multiple users is necessary to enable the robot to interact with each user according to person-alized social relations. If user identification takes a long time, it interferes the social interaction, so real time processing model, ARTMAP [18] is used in this module. Next, it recognizes the user's emotion from the facial expression. The reactions to user's emotions differ from each other even though the user says the same thing. In our system, the facial expression recognition is operated through the Google Cloud Vision API to see what the user's emotional state is.

The speech recognition module analyzes user's voice from the sensor and converts the speech sound data to a text format. The extracted text is sent to the language part to evaluate the meaning of the sentence user said. The user's touch also has a great influence on social interaction. The touchless interactions are not sufficient since there is a limitation of expression by just speaking. Touch recognition module distinguishes whether the touch input is patting or bullying from users. If the touch message con-tinuously occurs, looks like giving a pat, the module classifies the touch data as patting. This can happen when the user is satisfied to the robot. If the touch message repeatedly occurs in short time, looks like futzing badly, the module classifies the touch data as bullying which can happen when the user doesn't like the robot. These touches affect the emotion formation of the robot in the emotion generation and expression memory module of memory part.

2.3   Language Part

Since humans feel more natural and comfortable to communicate by voice rather than by text chatting, our framework is designed to communicate with natural voice con-versation. Language part analyzes meaning of the text detected by the speech recog-nition module to understand user's speech. As users express their inner side through speech, language part can realize the user's emotions, conditions, situations, and thoughts. To continue the voice conversation without interruption, the reply of the robot is also synthesized by voice and produces output.

The language part has two parts. First, the dialogue generator produces a sentence for the robot to answer in response to each user's speech sentence. Second, the text to speech synthesizer converts the answer text into sound data for the robot to speak out because users prefer to interact with robot spontaneously via voice. Details of both modules are described in APPENDIX.

2.4 Memory Part

In the framework, two kinds of memories are designed. One is working memory that manages data flows in the framework and makes decisions for tasks. The other is emotion-related memory to store emotional interaction experiences (emotional episodes) and recall expressions.

Working memory consists of modules for managing and scheduling multiple tasks. Each manager module determines what data to be used from the recognition, language, and memory part and what robot's action to take, so that the tasks are conducted properly without collision in the software. Decision is transmitted to the control part to move the body of the robot, and to the communication part which shows facial expression and sound to the person.

The emotion-related memory enables a robot to store interaction experiences, operates the emotional episode learning, and determines emotional expressions to users. Once the robot meets and interacts with a user again, it needs to take out the character of the user from the memory and build up a developed character by stacking the episodes. Our framework stores these emotional episodes in the memory part so that the robot can establish social relationship.

The emotion-related memory consists of an emotion generation and expression memory, Hierarchical Emotional Episodic Memory (HEEM) [17]. The emotion generation and expression memory lets the robot have its own emotion. Simply, it generates robot's bad emotion in response to user's teasing comments and bullying actions, or appearance of unfriendly user according to the episodic memory retrieval. It generates robot's positive emotion in response to user's compliments and patting actions, or appearance of friendly user. Then pre-trained sequences of the robot's neck and body gesture and facial expressions are recalled from the expression memory.

HEEM [17] is created by using deep Adaptive Resonance Theory (deep ART) [19, 20] for social human robot interaction. By storing and accumulating episodes with multi users in the HEEM, the relationship with each user can be reformed independently and constantly. HEEM enables to learn which emotions are correlated with past experiences of each user, predict upcoming episodes that could probably occur according to a user, and show proactive emotional reactions to the user, such as empathizing or repulsive reaction.

2.5 Communication Part

Communication part enables users to know lively of robot's reaction through voice, lip sync, and facial expression in order that users don't feel any sense of heterogeneity to robots. Instead of making a new robot head using dozens of motors, a virtual face is developed by face simulator. Since there are 50 muscles on the human face that facilitate about 7,000 different facial expressions, and it is hard to mount on the robot's head. The face simulator shows robot's facial appearance through a display screen which is reconstructed from a single 2D image [21], and shows the emotions sent by the facial expression module. It also has lip sync module that generates lip movement in the face simulator when the robot activates the sound player module to speak out the synthesized voice answer from language part. Lip sync movement and five fundamental expressions can be shown: anger, disgust, sadness, joy, and surprise (Fig. 2).

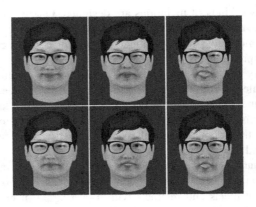

Fig. 2. Facial expressions of the robot. From the top left, the expressions are smile, disgust, anger, sad, surprise, and lip sync movement. It is possible to show each expression with five levels from the expressionless to the strongest expression.

2.6   Control Part

Control part has modules to control robot's neck movement and body motion. The neck movement controlled by the control part plays a decisive role in conversing with the user in the eyes by face tracking. Conducting dialogue with the eye contact empowers for more sincere and focused communication giving the user comfort and confidence. The gaze controller module controls robot's neck movement for eye contact with user or for expressing neck gestures. In order to make the robot keep eye contact with a user, the robot's neck moves so that the face of the person is centered in the camera image. Movements of the neck (panning and tilting) are controlled in proportion to the pixel distance between the center of the human face area and the center of the image. The smaller the face size, the smaller the width of motion assuming a smaller face means a farther face. In addition, as the pixel distance gets shorter, the speed of the neck movement is reduced to eliminate wobble. If there are several users in an image frame, robot's neck follows the average face position and size.

Answering the user with the appropriate body action to the user overcomes the insufficient interaction with words alone. Expressing emotions and thoughts as actions with words reaches the user more feelingly. The body gesture controller module controls robot's arm movement for expressing body gesture. Once the robot determines what action to take, the command for that action is sent to the controller in the Mybot's body platform through the TCP/IP network. Then, the controller operates the robot's motor movements based on ROS software. The arm and body behaviors are predefined sequences of movements, and in this paper, they are designed to express the robot's emotional gestures which are greetings, dislike, sorrow, joy, like, and surprise. The video is available at https://youtu.be/YVxyEEyGjLo.

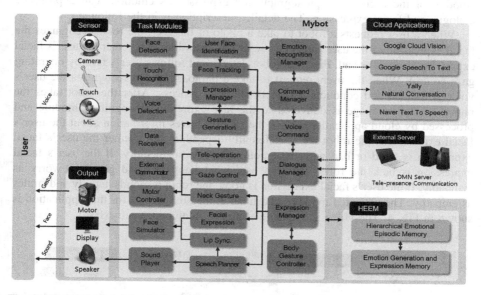

Fig. 3. Functional modules in the proposed framework. Later it will show at a glance which tasks are activated and which functional flows are made in the scenarios

3 Experiments

Scenarios are designed to demonstrate the effectiveness of the proposed human robot social interaction framework integrated in robotic platform. Additionally, we conducted a survey to evaluate how users feel about the proposed framework in an objective way. Figure 3 is a supplementary technical diagram which shows detailed connections between functional modules in our framework.

There are two scenarios. The first scenario, photo shooting with free talking, shows that the robot can recognize user's emotions and empathize with them. Second scenario shows the fundamentals of social relationships by showing how robot behaves differently when they see different users and depending on the emotions it feels from each user. Besides, the relation with one person is not set from the first episode, it can be changed by the further episodes.

3.1   Photo Shooting

This experiment was designed to prove whether the robot can recognize the user's emotions and respond appropriately. Using the proposed framework, the robot could perceive emotions of the users and react with proper dialogue and facial expression. In order to verify this capability, we conducted a scenario that the robot takes a picture with voice interaction. In the scenario, the robot automatically takes a picture when it recognizes a sentence that the user wants to take a picture. Then, the robot infers the emotion of the user in the photograph and expresses its emotion with appropriate dialogue and facial expression. The detailed description is in the following.

When a user requests a robot to take a photo of the user, the robot recognizes the facial expression of the user and shows reactions according to whether the user's facial expression is good or bad. First, the robot detects and follows user's face by operating gaze control. When the users request to take a photo, the robot recognizes the user's voice command and conducts the photo shooting process. After the robot takes the photo, the taken image is delivered to the face expression recognition module and processed by cloud application. When the robot receives information about the user's facial expressions, the robot reacts differently depending on whether the user's facial expression is good or bad with the emotion expression memory. If the user's facial expression is bad, the robot asks the user what happened and gives an worry as sympathy. If the user's facial expression is good, the robot responds happily and smiles as empathy. The activated modules and links in the architecture for this application is shown in Fig. 4. The video is available at https://youtu.be/BXpeLyxHst0.

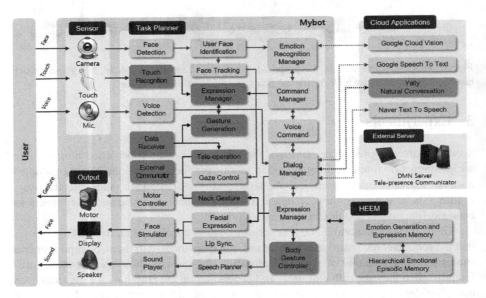

Fig. 4. Photo shooting scenario flow. The yellow boxes are the functional modules that are activated when this scenario is implemented (Color figure online)

3.2   User Identification and Robot's Differentiated Emotional Reactions

This scenario is designed to prove the advantages of the proposed emotional episodic memory. First, the robot identifies the users, feels different emotions to different users, and can provide appropriate services for each user. The interactions which include dialogues and touch sensing between the robot and the users generate emotions of the robot. If the user gives a positive sentence as 'you are handsome' or pats the robot, the robot feels friendliness, and in the opposite case, feels unfriendliness. Therefore, the robot can provide various services according to the emotions to each user. Second, the emotions for the users are not determined by a single interaction, but are progressively developed through past interaction experiences. Therefore, even if the robot feels unfriendly to the user, user can become a friendly user through various interactions. In the scenario, two users interacts with the robot several times, and we examined the responses of the robot. The robot could provide different services for each user, and the relationship with each user gradually changed. The activated modules and links in the architecture for this application are shown in Fig. 5. The video is available at https://youtu.be/LIe1yN_DjDk and the additional detailed description is in the following.

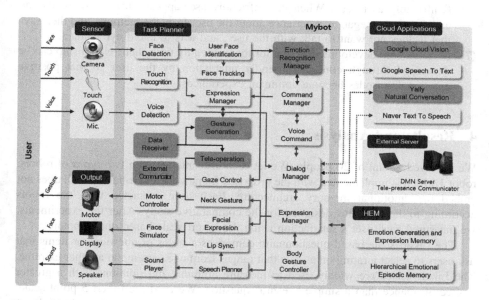

Fig. 5. User identification and robot's differentiated emotional reactions scenario flow. The yellow boxes are the functional modules that are activated when this scenario is implemented (Color figure online)

This application is an experiment that integrates the all of the proposed human robot social interaction framework. The scenario is as follows.

(1) (Unknown user registration) When a unknown user first appears to the robot, it asks who the user is, and when the user identifies his/her name, the robot registers the user's face with the name, and then it starts learning.

(2) (Normal user case) When a user appears before the robot and there is no special emotional experience with the user yet, the robot greets with normal gesture, nodding head (greetings).

(3) (Friendly user case) The user pats the robot and gives compliments to the robot. The robot has positive emotions. Then, the user information and positive emotional experience about the user are stored in episodes.

(4) (Friendly user case) When the friendly user appears, the robot recalls and anticipates the positive experience about the user from the episodes, and then it shows friendly greetings, in which the robot uses arm gestures and expresses a smiling face unlike normal greetings.

(5) (Unfriendly user case) Conversely, in this case, the user bothers the robot and makes negative comments about the robot. Then, the user information and negative emotional experience about the user are stored as emotional episodes in the memory.

(6) (Unfriendly user case) When the unfriendly user appears, the robot recalls and anticipates the negative experience about the user from the memory, and then says that it does not want to talk with the user, and acts to avoid eye contact.

(7) (Relationship development) Even for the unfriendly user, their relationship can be developed through continual friendly interaction by the user so that positive emotional episodes become dominant experiences.

4   Results and Analysis

We surveyed a total of 20 people. We let each user interact with two types of robots: robot 1 which is implemented with the proposed framework and robot 2 which does not embed the emotional memory architecture. Each user interacted with the robots according to the scenarios designed for the comparative study. After the experiments, we measured each user's satisfaction using the Likert scale based questionnaire.

The Likert scale based questionnaire comprises a series of questions or items that ask users to select a rating on a scale that ranges from one extreme to another extreme. The Likert scales are widely used to measure attitudes and opinions with a greater degree of nuance than a simple "yes/no" question. We used the Likert 5 point scale: very negative (1), negative (2), neutral (3), positive (4), very positive (5). Table 1 describes the questionnaire we used for the survey.

Questions asks whether the user could socially interact with the robot and feel intimacy from the robot. Furthermore, we asked naturalness, efficiency and satisfaction of the interaction with each robot. Especially for question 1, we made additional questions to distinguish which of the factors made it possible for users and robots to socially interact. We investigated three factors: face tracking, emotion recognition, and empathy ability.

Table 1. Questionnaire for the survey

No.	Questionnaire	Robot 1	Robot 2
1	It was possible to interact socially with the robot	3.6	1.55
2	(If you answered 4 or 5 point for No. 1) Face tracking was the factor that enabled social interaction	4.23	–
3	(If you answered 4 or 5 point for No. 1) The reason that social interaction was possible was because it recognized my emotion	4	–
4	(If you answered 4 or 5 point for No. 1) The reason that social interaction was possible was because the robot empathized with my emotions and laughed together	3.92	–
5	I felt intimacy with Robot	3.8	1.9
6	Do you think the robot is natural to build social relationships?	3.95	1.55
7	Do you think that the interaction of robot is effective in forming a social relationship?	4	1.85
8	Are you satisfied with the interaction of robots as social interaction?	3.95	1.7
9	For the proposed framework robot, Did it contributed to social interaction because it reacts differently to different users?	4.6	–
10	For the proposed framework robot, Did it contributed to social interaction because it develops the relationship, not fixing by a single episode?	4.26	–

Table 1 summarizes the experiment results. Robot 1 with the proposed interaction framework outperformed Robot 2 which does not contain the emotional memory unit and communication unit. Evaluated by Likert Scale, Robot 1 achieved exceedingly higher scores than Robot 2 in terms of possibility of social interaction, intimacy, naturalness, efficiency and satisfaction.

The survey takers responded that it was more possible to socially interact with Robot 1 than Robot 2. We also investigated the factor which made the social interaction possible through the following questions. As a factor that made social interaction possible was 4.23 for face tracking, 4.0 for emotion recognition, and 3.92 for robot's empathy ability. The biggest factor that made the social interaction possible was that the robot communicated with the user on eye-to-eye. Face tracking indicates that the conversation partner is paying attention. The experiment result shows that face tracking plays the most important role in social interaction with the users.

On the other hand, the overall naturalness of building social relation with Robot 1 is much higher than that of Robot 2 (factor of 2.55). This might have resulted from the framework's ability to identify each user, track face during conversation, express facial expressions and produce body gestures. These features enable the users to experience the conversational setting similar to the setting with real people. Another factor for natural relationship arises from adoption of HEEM. Robots with the proposed framework behaves differently to multi-users and the relationship improves over time without any fixation for one user. Since the highest score 4.6 is achieved in question 9, the factor that the robot responds individually to different users contributed the most to social interaction than the face tracking, emotion recognition, and the empathy.

The result for question 10 is 4.26 which is lower than for question 9, so we are planning to re-investigate by supplementing the scenarios for the relationship development part. In addition, the relationship now changes from unfriendly to friendly rapidly with just two actions: compliment and patting. More realistic and complex algorithm for the progressive improvement of the relationship is need to be studied for the thorough social interaction.

5   Discussion and Conclusions

Another contribution of this paper is that the whole system including perception, recognition, decision making, and execution is implemented and performs in several seconds. If the whole process took more than 10 s, it would be difficult to have natural interaction with human because human is prone to get bored with waiting. The computation time for each module is quantified and listed in Table 2. For space complexity, the proposed memory model has order of $O(nt + mn)$ where m, n, and t indicate the number of episodes, the number of unique events, and the input dimension, respectively. Supposing that an input vector has 1,000 elements and uses 1 Bytes for each element, theoretically, 3 GByte memory can contain around 50,000 unique events and 50,000 unique episodes. Of course, these numbers can be traded off under conditions that satisfy the space complexity, and they are very enough to cover practical experiment situation. The total memory usage including all background software of Microsoft Windows is around 4 GBytes out of 8 GBytes. Thus, typically, the proposed framework doesn't have significant problems in this computing environment.

Table 2. Computation time for each module

Module	Time (seconds)	Unit
Face detection (OpenCV, Haar Casecades)	~0.1	per 640 * 360 image frame
Face identification (ARTMAP)	<0.001	per 24 * 24 image
Facial expression recognition* (Google cloud vision)	~3	per image frame
Speech recognition* (Google)	~2	per sentence
Dialog generation* (Yally)	~0.5	per sentence
Text to speech synthesis* (Naver)	~1	per sentence
Hierarchical emotional episodic memory learning and retrieval	<0.001	per event
Emotion generation and expression memory retrieval	<0.001	per event

*Computation time for cloud applications are depending on internet speed and cloud server condition. The wireless device equipped in our system has 780 Mbps.

Our proposed framework enables the robot to interact more naturally with the users by using the emotional episodic memory. The emotional episodic memory stores the interaction experiences with each user and the entailing emotions. Thus, the robot with the proposed framework could provide personalized services. In addition, since the emotions generated by the framework gradually developed through interaction experiences, the robot could form natural social relationships with users. Finally, we implemented the proposed framework in an interactive robotic platform named, Mybot that operates in real time. The effectiveness and applicability of the proposed framework are demonstrated by showing two application scenarios with user study.

At present, interactions between the robot with the proposed framework and users are based on autonomous dialogue, a few touches, and visual images. Therefore, if the framework adopts the Visual Question Answering (VQA) technology, it enables the robot to interact with users in more various ways by utilizing the images of surrounding environment. We are working on building an interactive VQA system using our proposed framework [22]. Additionally, if the performance of each cloud application gets better, we expect our framework to be more practical and meticulous.

Acknowledgement. This work was supported by the National Research Foundation of Korea (NRF) grant funded by the Korea government (MSIT) (No. NRF-2017R1A2A1A17069837).

Appendix: Hardware and Software Implementation Details

For the hardware design, we developed a humanoid type robotic platform aiming at sufficient interaction with humans. The hardware robot is named Mybot, developed in the RIT Laboratory at KAIST (Fig. 6).

Mybot has an upper body including two arms (10DoFs for each arm) and one trunk (2DoFs), and a lower body including an omnidirectional wheel and power supply as shown in Fig. 6. It is running on Linux 16.04 operating system and controlled by an Odroid. The body is connected to the robotic head via TCP/IP communication.

Tablet PC was used for robotic head which function as image receiver, voice detection, touch detection in our experiment since it has various input sensors and output interfaces. Especially for the touch detection module, the touch sensor on the tablet PC recognizes mouse clicks and mouse movements as the touch input, mouse movements are tightly restricted during the experiment.

A tablet computer with Windows10 64bits OS, i5 6th CPU, 8G RAM is used, and it has 12.3″ 2763 * 1824 resolution display with touchable screen. The device also has a 5 M pixels front camera, a microphone, and a speaker equipped.

For the neck frame, 3 actuators are used for 3 DOFs motion: pan, tilt, and yaw of the robotic head. The actuators are ROBOTIS MX-64R motors that operate at 15 V and have around 80 RPM speed and 80 kg * cm stall torque. The actuators are connected to the robotic head (tablet) via USB to Dynamixel interface.

Fig. 6. Mybot, the interactive robotic hardware platform.

For the software design, our team used visual C++ MFC programming to implement the proposed framework. As we had to use some cloud application in recognition part and language part, we used socket server and socket client to access to the API. We used Internet access to 4 cloud applications to give the framework functionality.

Face detection module uses OpenCV libraries with the Haar Cascades method. The user face identification module classifies user's face so the robot can identifies which user is interacting with. For the identification algorithm, ARTMAP is applied, which is a supervised learning version of Adaptive Resonance Theory (ART) network [18]. The reason for applying ARTMAP is that facial learning and recognition should be conducted in real time, and feasible performance can be achieved even with small number of samples. More technically, the robot takes a 640×360 image, and crops and resizes the image in the range where the user's face is located. Then, the image is vectorized in one dimensional vector, and used as the input vector of ARTMAP. The result video of the user identification in real time is available at https://youtu.be/Ik_FwL2WYK8.

The facial expression recognition module recognizes user's facial expression. The module uses Google Cloud Vision which provides recognition API for four human emotions: joy, sorrow, anger, and surprise; and with four levels: very unlikely, unlikely, likely, and very likely. The advantage of Google Cloud Vision is that it is available for any user face.

Speech recognition module uses Google Speech to Text cloud application which has the function of converting the speech of a user into a text file. Additionally, it supports multi-lingual recognition services including Korean language service with the state-of-the-art performance.

Social relationships cannot develop or sustain without daily conversation. Thus, dialogue generator module takes text data from the speech recognition module and delivers the text to Yally's Natural Conversation cloud application (http://www.yally.com/en/), and gets answer text from it. The generated answers are everyday life conversation rather than specific conversation.

Text to speech synthesizer module uses Naver Text to Speech cloud application (Clova Speech Synthesis). This module is directly linked to the Lip Sync. module in the communication part, so it signals when the lip synchronization should start.

References

1. Lin, P., Abney, K., Bekey, G.A.: Robot Ethics: The Ethical and Social Implications of Robotics. The MIT Press, Cambridge (2014)
2. Scheutz, M.: What is robot ethics? [TC spotlight]. IEEE Robot. Autom. Mag. 20(4), 20–165 (2013)
3. Fong, T., Nourbakhsh, I., Dautenhahn, K.: A survey of socially interactive robots. Robot. Auton. Syst. 42(3–4), 143–166 (2003)
4. Knight, H.: How Humans Respond to Robots: Building Public Policy Through Good Design. Brookings, Washington, DC (2014)
5. Bartneck, C., Forlizzi, J.: A design-centred framework for social human-robot interaction. In: 13th IEEE International Workshop on Robot and Human Interactive Communication 2004, ROMAN 2004. IEEE (2004)
6. Belpaeme, T., et al.: Multimodal child-robot interaction: building social bonds. J. Hum. Robot Interact. 1(2), 33–53 (2013)
7. Gorostiza, J.F., et al.: Multimodal human-robot interaction framework for a personal robot. In: The 15th IEEE International Symposium on Robot and Human Interactive Communication 2006, ROMAN 2006. IEEE (2006)
8. Glas, D., et al.: An interaction design framework for social robots. Robot. Sci. Syst. 7, 89 (2012)
9. Duffy, B.R., Dragone, M., O'Hare, G.M.P.: Social robot architecture: a framework for explicit social interaction. In: Android Science: Towards Social Mechanisms, CogSci 2005 Workshop, Stresa, Italy (2005)
10. Breazeal, C.L.: Designing Sociable Robots with CDROM. MIT Press, Cambridge (2004)
11. Złotowski, J., Strasser, E., Bartneck, C.: Dimensions of anthropomorphism: from humanness to humanlikeness. In: Proceedings of the 2014 ACM/IEEE International Conference on Human-Robot Interaction. ACM (2014)
12. Arkin, R.C., et al.: An ethological and emotional basis for human–robot interaction. Robot. Auton. Syst. 42(3–4), 191–201 (2003)
13. Kim, H.-R., Lee, K.W., Kwon, D.-S.: Emotional interaction model for a service robot. In: IEEE International Workshop on Robot and Human Interactive Communication, ROMAN 2005. IEEE (2005)
14. Lee, W.H., et al.: Motivational emotion generation and behavior selection based on emotional experiences for social robots. In: Workshops in ICSR 2014 (2014)
15. Miwa, H., et al.: A new mental model for humanoid robots for human friendly communication introduction of learning system, mood vector and second order equations of emotion. In: IEEE International Conference on Robotics and Automation, ICRA 2003, vol. 3. IEEE (2003)

16. Kwon, D.-S., et al.: Emotion interaction system for a service robot. In: The 16th IEEE International Symposium on Robot and Human interactive Communication, RO-MAN 2007. IEEE (2007)
17. Lee, W.-H., Kim, J.-H.: Hierarchical emotional episodic memory for social human robot collaboration. Auton. Robots **42**(5), 1087–1102 (2018)
18. Carpenter, G.A., et al.: Fuzzy ARTMAP: a neural network architecture for incremental supervised learning of analog multidimensional maps. IEEE Trans. Neural Netw. **3**(5), 698–713 (1992)
19. Park, G.-M., Kim, J.-H.: Deep adaptive resonance theory for learning biologically inspired episodic memory. In: 2016 International Joint Conference on Neural Networks (IJCNN). IEEE (2016)
20. Park, G.-M., et al.: Deep art neural model for biologically inspired episodic memory and its application to task performance of robots. IEEE Trans. Cybern. **48**(6), 1786–1799 (2018)
21. Yun, J., et al.: Cost-efficient 3D face reconstruction from a single 2D image. In: 2017 19th International Conference on Advanced Communication Technology (ICACT). IEEE (2017)
22. Cho, S., Lee, W.-H., Kim, J.-H.: Implementation of human-robot VQA interaction system with dynamic memory networks. In: 2017 IEEE International Conference on Systems, Man, and Cybernetics (SMC). IEEE (2017)

Convolutional Neural Network-Based Collaborative Filtering for Recommendation Systems

Yat Hong Low$^{(\boxtimes)}$, Wun-She Yap, and Yee Kai Tee

Lee Kong Chian Faculty of Engineering and Science,
Universiti Tunku Abdul Rahman, Petaling Jaya, Malaysia
lowyathong@outlook.com, {yapws, teeyk}@utar.edu.my

Abstract. Collaborative filtering based recommender systems that predict the user preference based on their past interactions have been adopted by many online services. Matrix factorization that projects users and items into a shared latent space is one of the popular collaborative filtering techniques. Recently, a general neural network-based collaborative filtering (NCF) framework, employing generalized matrix factorization and multi-layer perceptron models termed as neural matrix factorization (NeuMF), was proposed for recommendation. Meanwhile, convolutional neural network (CNN) is a variation of a multi-layer perceptron commonly used in computer vision. CNN is also normally used to model user profiles and item descriptions for recommendation. In this work, the CNN was used differently, that is, to model the interaction between user and item features directly in the recommendation systems. More specifically, a special case of NCF that employs matrix factorization and CNN was proposed. The model used general matrix factorization to model latent feature interactions using a linear kernel and the CNN to learn the interaction function from data using a non-linear kernel. Experiments conducted on a public dataset, Movielens, demonstrated that the proposed model was superior when compared to the published NCF framework such as NeuMF and other state-of-the-art methods.

Keywords: Recommender system · Machine learning ·
Convolutional Neural Network · Collaborative filtering

1 Introduction

In the current age of the Internet, roughly 51.7% of the world population are internet users [1]. The most common activities that the internet users do when online are reading articles, watching videos, listening to music and making online purchases. In order to improve user experience, customer satisfactions and to avoid information overload, recommender systems have been developed to recommend products/services that are closely relevant to the users' interest or preference. Through the competitions such as Recsys Challenge [2] and Netflix Prize [3], many different types of recommender systems have been proposed and developed.

© Springer Nature Singapore Pte Ltd. 2019
J.-H. Kim et al. (Eds.): RiTA 2018, CCIS 1015, pp. 117–131, 2019.
https://doi.org/10.1007/978-981-13-7780-8_10

In general, recommender systems are split into two major categories [4]: content based and collaborative filtering. The former works by creating a profile for each user and/or product. For instant, user profiles may content demographic information and product profiles consist of synopsis and participating actors, assuming it is a movie profile. Basically, keywords are used to describe every product, users are then given recommendations based on the keywords associated with the products they purchased in the past or popularity of the item among their demographic group. Nevertheless, content based recommender systems require to gather external information such as demographic information of the users which may not be available or easy to collect.

In the collaborative filtering method, a recommendation is based on the past interaction between the user and the product, where information such as user or product profiles are not required. In other words, the recommender system will not know about the details of the products when a recommendation is suggested yet it can recommend the product that can be converted to sales (higher accuracies/chances) compared to content based methods [4]. Among the collaborative filtering techniques, matrix factorization method [5] is the most popular. The method works by first representing every user and item with a vector of latent features and then projecting the user and item latent features into latent vectors. As a result, the user-item interaction can be approximated by the inner products of the latent vectors.

Various efforts have been made to improve the collaborative filtering method using matrix factorization such as merging the neighbourhood model with the matrix factorization methods [6], using factorization machines for a generic modelling of features [7] and combining matrix factorization with topic models of item content [8]. Although matrix factorization is the most popular technique for collaborative filtering, its performance is highly dependent on the choice of the interactive function – inner product, where the performance can be improved by introducing user and item bias terms into the inner product [9]. However, simply multiplying the latent features linearly may not be sufficient to describe the complex interaction of the user and the item. Another common issue associated with a recommender system that leads to less efficient performance of the existing methods in the literature is the large sparsity of the complex interaction of the user and the product, many zero elements in the user-item interaction matrix.

Recently, deep neural networks (DNNs) which have achieved remarkable success in computer vision and natural language processing have been proposed to address some of the shortfalls discussed to improve collaborative filtering. For example, Salakhutdinov et al. [10] applied a class of two layer undirected graphical models named Restricted Boltzmann Machines, a type of generative stochastic neural network that learns a probability distribution over its set of inputs, to model explicit item ratings provided by users; it was shown that the performance was 6% better than the Netflix's own system scores which contain over 100 million ratings and large sparsity. He et al. [9] integrated Multi Layer Perceptrons (MLP) and matrix factorization into a neural network-based collaborative filtering (NCF) framework to model user and item latent vectors that could learn an arbitrary function from data to overcome the complex interaction of the user and the item; the proposed framework showed significant improvements over other published methods on two real-world datasets, MovieLens and Pinterest. However, every neuron on one layer is connected to every neuron in the next layer in MLP, this dense connection makes the model rigid when it comes to feature learning, giving it difficulties to learn new features.

Convolutional Neural Network (CNN) is a feedforward neural network originally developed for computer vision. While not commonly applied to the field of recommender systems, there were studies on the usage of CNN to extract auxiliary information in order to create recommendations. CNN had been used in music recommendation by analyzing acoustic of the songs and making the predictions based on the item latent model approximated [11], and video recommendations by modelling item latent factors based on the reviews for videos and integrated it with Probabilistic Matrix Factorization [12].

In this work, CNN was used to directly model the user-item interaction from data and integrated it with matrix factorization to provide recommendations. CNN was explored because there is a convolution layer that extracts local features by convolving input signals from adjacent neurons, allowing it to be more flexible when learning features. The proposed model was compared with the state-of-the art methods to investigate which method is more suitable to model the complex user-item interaction.

2   Preliminary

In this section, the implicit feedback of recommender systems will be discussed, followed by matrix factorization.

2.1   Implicit Feedback

Recommender systems rely on different types of input to generate a recommendation, the most useful input is the high quality explicit feedback provided by the users regarding their interest in products such as star ratings or thumbs-up/down/like button. However, this kind information may not be always available, leading to the need to use the more abundant implicit feedbacks such as purchase history, browsing history, search patterns and mouse movement which may indirectly reflect the user preference. The work in this study focused on implicit feedback because they can be tracked automatically. Although implicit feedbacks are easier to collect, it is more challenging to use because they are not a direct reflection of user preference and there is usually hardly any negative feedback (difficult to link whether the users dislike the items).

Assuming U and I represent the number of users and items respectively, the user-item interaction matrix, $P = [p_{ui}]^{|U| \times |I|}$, whose elements p_{ui} represents the interaction between user u and item i where,

$$p_{ui} = \begin{cases} 1, & \text{if an interation is observed;} \\ 0, & \text{otherwise.} \end{cases} \tag{1}$$

The user-item interaction, p_{ui} will have the value of 1 when there is an interaction and 0 when there is not. It should be noted when p_{ui} is 1, it does not mean that user u actually likes item i, it merely shows that user u is aware of the item and interacted with it, likewise, when p_{ui} is 0, it does not mean that the user does not like the item, it could be that the user is unaware of the item, causing an unobserved entry in the user-item interaction matrix. This reflects the challenges in using implicit data for recommendations; the observed entries (1) indicate the users are aware of the items but there are many unobserved entries (0), where these can be just missing data or the item list is too long, the users are not able to browse through all the items.

The goal of recommendation with implicit feedback is to generate a ranked list of item(s) that reflects the users' preference by estimating the scores of unobserved entries in the user-item interaction matrix, P. In general, model-based approaches assume that the user-item interaction can be described as $\hat{p}_{ui} = f(u, i|\Theta)$, where $\hat{p}_{ui}$ refers to the estimated score of interaction p_{ui}, f is the model/function used to generate the estimated score and Θ denotes the model parameters of f.

There are few ways to optimize the model parameters, Θ in order to obtain the best estimated score of interaction. The most commonly used methods are pointwise loss [4, 13] and pairwise loss [14, 15] objective functions. The previous usually tries to minimize the squared loss between $\hat{p}_{ui}$ and p_{ui}, and treats all the unobserved entries as negative feedback or selectively choose some unobserved entries as negative [13]. For the pairwise loss method [14], the observed entries are ranked higher than the unobserved entries thus it maximizes the margin between the observed and unobserved entry instead of minimizing the loss between $\hat{p}_{ui}$ and p_{ui}.

In this study, a neural network was used to estimate $\hat{p}_{ui}$ leading to the support of both pointwise and pairwise learning. More details are available in the later section.

2.2 Matrix Factorization

Conventional collaborative filtering techniques are often classified into two categories: neighbourhood models [16, 17] that recommend items based on similarity measures and latent factor models [6] that recommend items based on the latent factors discovered by the model. From various studies in the literature, the latent factor models tend to have higher accuracy when it comes to generating recommendations when compared to the previous category [6] and matrix factorization is one of the most popular latent factor models.

In matrix factorization, each user and item are described with a real-valued vector of latent features. Assuming x_u and y_i refer to the latent vector for user u and item i respectively, an estimated score of interaction using matrix factorization can be obtained by the inner product of x_u and y_i:

$$\hat{p}_{ui} = f(u, i \mid x_u, y_i) = x_u^T y_i = \sum_{d=1}^{D} x_{ud} y_{id}, \tag{2}$$

where D is the dimension of the latent space.

When matrix factorization is used to model the interaction of user and item latent factors, it hypothesizes each dimension of the latent space is independent of each other and they can be linearly combined with the same weight, in another word, a linear model of latent factors. However, a simple and linear inner product of x_u and y_i is inadequate to estimate the complex user-item interactions in low dimensional latent space. One of the possible solutions is to increase the number of latent factors but it may suffer the risk of over fitting the data, especially when the sparsity is high.

3 Convolutional Neural Network-Based Collaborative Filtering

In this section, the details of the proposed model, convolutional neural network-based collaborative filtering inspired by the NCF framework [9], are discussed. Firstly, the general framework of the model will be presented, followed by the description of the individual components of the proposed model: General Matrix Factorization, Convolutional Neural Network and its layers, the ensemble between the two components and lastly, the objective function used in the training to optimize the model parameters.

3.1 General Framework

Figure 1 shows the general framework of collaborative filtering where a multi-layer representation was used to model the user-item interaction, p_{ui}. The first layer is the input layer, where the one-hot encoded IDs of user u and item i were supplied as the input. Both the user and item IDs were then fed into the embedding layer; this is a fully connected layer that maps the sparse representation to a dense vector. The user and item embeddings were the latent vectors in the latent factor model discussed above and they were then passed to the collaborative filtering layers to model the user-item interaction by mapping the latent vectors to the prediction scores. Each layer in the collaborative filtering layers could be customised to model certain latent structures of user-item interactions. At the output layer, the predicted score was obtained after training the framework by minimizing the prediction ($\hat{p}_{ui}$) and target (p_{ui}).

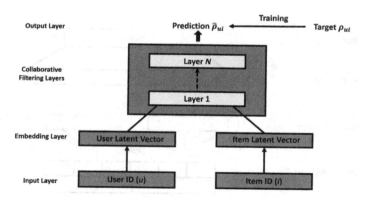

Fig. 1. General collaborative filtering framework

3.2 Generalized Matrix Factorization (GMF)

As mentioned, matrix factorization is one of the most popular collaborative filtering methods for recommender systems. It has a framework as shown in Fig. 1 too. One hot-encoding of user and item ID were generated in the input layer and passed to the embedding layer to obtain the latent vector of users, x_u, and items, y_i. After that, the element-wise product of x_u and y_i were calculated in the collaborative filtering layer as below:

$$z^{gmf} = \varphi_1(x_u, y_i) = x_u \cdot y_i \tag{3}$$

Lastly, the prediction in the output layer could be obtained using the following equation:

$$\hat{p}_{ui} = \sigma\left[h^T z^{gmf}\right]$$

where σ is the sigmoid function given by $\sigma(a) = 1/(1 + e^{-a})$ and h is the weights of the output layer.

3.3 Convolutional Neural Network (CNN)

CNN is a type of feed-forward neural network originally developed for computer vision and later found useful in various other areas such as speech and text processing. In this work, the CNN was used differently, that is, to learn the interaction between user and item features with a large level of flexibility and non-linearity, unlike the GMF that uses only a fixed element-wise product of user and item features. Figure 2 shows the general architecture of a CNN which consists of input, embedding, convolution, pooling and output layer.

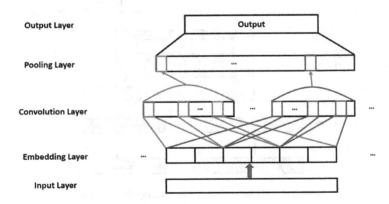

Fig. 2. General architecture of a CNN

Input

The first layer is the input layer, where two feature vectors describing user u and item I were used as input. Since the main focus of this study was on pure collaborative filtering setting, IDs of user u and item i were supplied and converted to a binarized sparse vector with one-hot encoding.

Embedding

In the second layer, the one-hot encoded user ID and item ID were fed to obtain their respective k-dimensional embeddings. These embeddings served as the latent vectors, x_u and y_i. The two vectors were then concatenated together to form the interaction vector, z^{cnn},

$$z^{cnn} = \varphi_1(x_u, y_i) = \begin{bmatrix} x_u \\ y_i \end{bmatrix}.$$

Convolution

The convolution layer was used to extract contextual features. The convolution architecture described in [18] was implemented to analyze the contextual features between the user and item latent vectors by processing the interaction vector, z^{cnn}, from the previous layer. For example, a contextual feature, c_n^m was extracted by the mth shared weight, $W_c^m \in \mathbf{R}^{2k \times w}$, whose window size w determines the number of surrounding factors,

$$c_n^m = f\left(W_c^m * z^{cnn}_{(:,n:(n+w-1))} + b_c^m\right), \tag{4}$$

where $*$ is a convolution operator, $b_c^m \in \mathbf{R}$ is a bias for W_c^m and f is an activation function. In this study, Rectifier Linear Unit was used as the activation function because it could avoid the vanishing gradient problem which normally leads to slow optimization and poor local minimum [19]. The contextual feature vector, c^m of an interaction with W_c^m could be the constructed as below:

$$c^m = \left[c_1^m, c_2^m, \ldots \ldots, c_n^m, c_{k-w+1}^m\right]. \tag{5}$$

It should be noted that one shared weight captures only one type of contextual features. In this study, multiple shared weights were deployed to capture multiple types of features and generate as many as j numbers of contextual feature vectors, so for example, if we chose to have 10 different shared weights W_c^m (where $m = 1, 2, 3, \ldots,$ 10), we will get 10 different contextual feature vectors.

Pooling

This layers extracts the representative features from the convolutional layer, where an interaction could be represented as j numbers of contextual feature vectors. From Eq. 5, there might be too many contextual features and most of them might not be useful.

Therefore, with the help of max-pooling, only the maximum contextual feature from each contextual feature vector was extracted.

$$z_f^{cnn} = \left[\max\left(c^1\right), \max\left(c^2\right), \dots, \max\left(c^j\right) \right] \tag{6}$$

where c^m is a contextual feature vector extracted by the mth shared weight, W_c^m.

In the event of multiple layers of CNN were being used, the vector from Eq. 6 would act as the input for the next layers, therefore:

$$\begin{aligned}
z_1^{cnn} &= \varphi_1\left(x_u, y_i\right), \\
z_2^{cnn} &= \varphi_2\left(z_1^{cnn}\right), \\
&\ \ \vdots \\
z_f^{cnn} &= \varphi_f\left(z_f^{cnn}\right) = cnn\left(W, \begin{bmatrix} x_u \\ y_i \end{bmatrix}\right),
\end{aligned} \tag{7}$$

where W denotes all the weight and bias variables to prevent clutter and $\begin{bmatrix} x_u \\ y_i \end{bmatrix}$ denotes the original interaction vector.

Output

At the output layer, the features obtained from the previous layer are then projected into the output layer to produce the estimated score:

$$\hat{p}_{ui} = \sigma\left[h^T z_f^{cnn}\right]$$

where σ is the sigmoid function given by $\sigma(a) = 1/(1 + e^{-a})$ and h is the weights of the output layer.

3.4   Hybrid of GMF and CNN

In the previous sections, two different recommender systems were introduced – GMF which uses a linear kernel to model the latent feature interactions, and CNN that applies a non-linear kernel to learn the interaction function from data. In this section, the proposed hybrid framework was introduced to better model the complex user-item interactions.

The simplest way to fuse the GMF and CNN was to let them shared the same embedding layer and combined the outputs of their interaction functions. However, sharing embeddings of GMF and CNN might degrade the performance of the hybrid model especially when the optimal embedding size of the two models were vastly different [9]. Thus, separate embedding layers were introduced to provide more

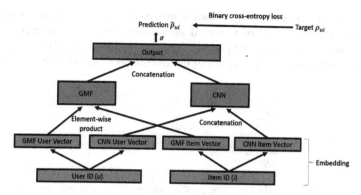

Fig. 3. Architecture of the proposed hybrid model

flexibility to the hybrid model and only combine the two models by concatenating their last respective layer as shown in Fig. 3.

The estimated scores/predictions of the hybrid model could be calculated as below:

$$z^{gmf} = x_u^{gmf} \cdot y_i^{gmf},$$

$$z_f^{cnn} = cnn\left(W, \frac{x_u^{cnn}}{y_i^{cnn}}\right),$$

$$\hat{p}_{ui} = \sigma\left(h^T \begin{bmatrix} z^{gmf} \\ z_f^{cnn} \end{bmatrix}\right), \qquad (8)$$

where x_u^r and y_i^r denotes the user and item embedding respectively for GMF if $r = gmf$ and CNN if $r = cnn$. We named this hybrid model as convolutional neural network-based collaborative filtering (CNNCF).

3.5 Objective Function

The implicit feedback recommendation in the study was converted to a binary classification problem, where 1 denotes the user has interacted with the item and 0 otherwise. However, having just the binary output, either 1 or 0, was a trivial problem. Therefore, a probabilistic approach was used in this study, where the prediction score, $\hat{p}_{ui}$, represents the likehood of i being relevant to u would be generated instead. To express the score in a probabilistic manner, the score had to be constrained in the range between 0 and 1, this was done by applying a logistic function on the activation function of the output layer. With this, the likelihood function is defined as:

$$p(O, O^- | X, Y, \Theta_f) = \sum_{(u,i) \in O} \hat{p}_{ui} \sum_{(u,i) \in O^-} (1 - \hat{p}_{ui}), \qquad (9)$$

where X and Y represents the latent factor matrix for users and items, respectively; O and O^- represents the set of observed interactions and set of negative instances, respectively; and Θ_f is the model parameter of the likelihood function.

By applying the negative logarithm of the likelihood function above, the following was obtained:

$$
\begin{aligned}
L &= -\sum_{(u,i)\in O} \log \hat{p}_{ui} - \sum_{(u,j)\in O^-} \log\left(1-\hat{p}_{uj}\right) \\
&= -\sum_{(u,i)\in O\cup O^-} p_{ui}\log \hat{p}_{ui} + (1-p_{ui})\log(1-\hat{p}_{ui})
\end{aligned}
\tag{10}
$$

L is the objective function known as the binary cross-entropy loss and was to be minimized during training using stochastic gradient descent [20] to produce the best predictions.

4   Experiments

In this section, experiments were performed to investigate the following research questions (RQs):

RQ1.   Is the proposed CCNCF better than the state-of-the-art implicit collaborative filtering methods?

RQ2.   How does the proposed optimization framework perform in the Top-K recommendation task?

RQ3.   Are more layers of CNN useful for learning the complex user-item interaction data?

4.1   Experiment Settings

Dataset

The performance of CNNCF was evaluated using the MovieLens - 1 M dataset [21]. The dataset consists of user ratings of movies (explicit feedbacks) but these were changed to implicit feedbacks by changing the entries to 1 if the user rated the item regardless of good or bad and 0 when there was no rating. The details of the MovieLens dataset used in this study is summarized in Table 1.

Table 1. MovieLens data statistics

Dataset	Number of interactions	Number of users	Number of items	Sparsity
Movielens	1000209	3706	6060	95.53%

Evaluation

The performances of the models used in this study were evaluated using the leave-one-out evaluation method [13, 14, 22]. This evaluation method was carried out by using the latest rating record of each user as the testing set and the remaining ratings as the training set. In order to save time, rather than ranking all of the items for every user when evaluating the performance, 100 items that were not interacted by the user were randomly selected and ranked following the evaluation protocols used in [23].

The metrics used in the study to measure the model performance were the Hit Ratio (HR) which measures if the test item is among the top-K list generated by the recommender and the Normalized Discounted Cumulative Gain (NDCG) that takes the position of the items in the list into consideration by awarding higher scores to higher ranking hits. The average score of both metrics for each user were calculated to check the performance of the recommendations.

Baselines and model implementation

Our proposed model are compared with the following competitive baselines:

- **NeuMF** [9]. This is the state-of-the-art method for item recommendation. A hybrid model where a matrix factorization model is coupled with a MLP model to generate the recommendations. It should be noted that the results used for comparison in this study were the NeuMF without pre-training.
- **BPR** [14]. A model that optimizes the matrix factorization model with a pairwise ranking loss, tailored to learn from implicit feedback.
- **ItemKNN** [17]. An item-based collaborative filtering method that computes the cosine of item to item similarity for recommendations.
- **GMF** [9]. A generalized latent factor model proposed within the NCF framework.
- **eALS** [13]. A matrix factorization method that optimizes using squared loss by treating all unobserved interactions as negative instances and weighting them non-uniformly based on the item popularity.

The models used in the study (GMF, CNN and CNNCF) were learnt by optimizing the binary cross-entropy loss (Eq. 10). The parameters of the models were randomly initialized using a Gaussian distribution with a mean of 0 and standard deviation of 0.01 followed by optimization via mini-batch Adam [24]. The learning rate of the models were set to 0.001 while the batch size was set to be 256. Since the number of latent factors used by the models determines the capabilities of the model, the effect of latent factors of (8, 16, 32, 64) on the model performance was evaluated. For example, a latent factor of 8 means we will set the embedding size of each IDs to 8, this is the same for the other latent factor numbers.

4.2   Results and Analysis

4.2.1   Performance Comparison with Baseline Models (RQ1)

Figure 4 shows the HR@10 and NDCG@10 for the proposed model and the baseline models with varying latent vector factors. The proposed CNNCF performed better than all the baseline methods in terms of HR and NDCG until 32 factors (the best results were obtained by CNNCF, HR@10 was 0.7068 and NDCG@10 was 0.4259) and had a slight drop in performance when the number of factors was increased to 64 potentially due to overfitting. Below 32 factors, our proposed method performed better than the state-of-the-art NeuMF, because the MLP in the NeuMF has every neuron of a preceding layer connected to every element of the next layer, resulting in a dense connection. When a feature that the neurons are supposed to be sensitive to changes, all the neurons will have to relearn the new feature, making learning harder. As for CNN, feature learning is much more flexible as the convolution layers help to extract the local features by convolving input signals with the adjacent neurons, leading to easier learning. In this performance comparison, CNNCF was kept at three layers because the MLP in NeuMF used three layers as well.

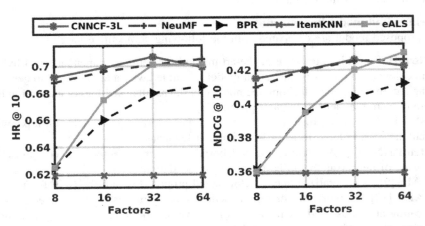

Fig. 4. Performance of HR@10 and NDCG@10 versus the number of factors for different models

4.2.2 Top-K Recommendation of CNNCF-3L and Comparison with GMF (RQ2)

Next, the performance of the Top-K recommended list produced by CNNCF-3L and GMF by fixing the factor at 16 and varying the value of K from 2 to 10 was compared. The purpose of this experiment is to study whether adding the CNN layers to the GMF will improve the recommendation performance. From Table 2, it can be seen that the performance of both the models improved as the K increased when measured with HR and NDCG and our proposed CNNCF-3L outperformed the most popular collaborative filtering method for all the tested K values. The best obtained results of our proposed model were 0.6988 for HR@10 and 0.4201 for NDCG@10. The improvement was more prominent when measured in term of NDCG which is more stringent because higher scores were awarded for higher ranking hits (around an average of 2.78%

Table 2. HR@K and NDCG@K with respect to the different values of Top-K recommendation list

Top-K	HR@K		NDCG@K	
	GMF	CNNCF-3L	GMF	CNNCF-3L
2	0.307	**0.3139**	0.2646	**0.2683**
4	0.4621	**0.4712**	0.3323	**0.3446**
6	0.5538	**0.57**	0.3664	**0.3779**
8	0.6328	**0.6434**	0.3906	**0.4035**
10	0.693	**0.6988**	0.4105	**0.4201**

improvement across all tested Top-K value after adding CNN layers to GMF). This implies that GMF which uses a linear kernel was not able to properly model the complex user-item interaction but our proposed hybrid model which includes CNN as part of the model was able to do a better job.

4.2.3 Effects of the Number of Layers in the CNN (RQ3)

Since CNNs are rarely used in learning the user-item interaction function directly, this section is performed to assess whether incorporating more layers of CNN is beneficial to the recommendation capabilities of our proposed model. Figure 5 shows the number of CNN layers versus the performance of HR@10 and NDCG@10 for 8, 16, 32 and 64 factors.

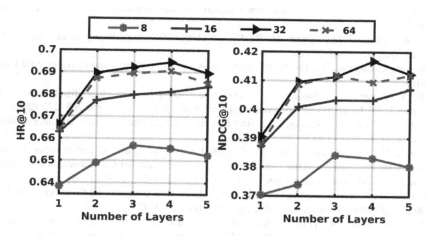

Fig. 5. HR@10 and NDCG@10 w.r.t the number of CNN layers

From the results in Fig. 5, as the number of layers in the CNN increases from 1 to 3, the performance of the model measured in term of HR and NDCG increases in general. However, increasing the layer further to 4 layers and beyond, the mixed results were observed which might be caused by overfitting by the CNN. The best obtained results were 0.6945 for HR@10 and 0.4168 for NDCG@10 with 32 factors and 4 layers of CNN.

5 Conclusion and Future Work

In this work, a recommender system which incorporates GMF to model latent feature interactions using a linear kernel and CNN which learns the interaction function from any data with the use of a non-linear kernel, was proposed. Several experiments were conducted on the real world dataset, MovieLens-1 m and it was shown that our proposed CNNCF model was able to outperform all the state-of-the-art methods in majority of the cases where the best performance was 0.7068 for HR@10 and 0.4259 for NDCG@10 with 32 factors and 3 layers of CNN.

In the future, an additional dropout layer will be added to the CNN to mitigate the overfitting problem and to include auxiliary information [25] such as user details, item reviews or may be even images into the model to enhance the user/item embedding in order to improve the performance of the proposed model.

Acknowledgement. We gratefully acknowledge Jin Zhe for sharing his computing resource (i.e. the Titan Xp GPU donated by the NVIDIA Corporation) to support this research at the preliminary stage. This research was also supported by the Collaborative Agreement with NextLabs (Malaysia) Sdn Bhd (Project title: Advanced and Context-Aware Text/Media Analytics for Data Classification).

References

1. Internet Usage Statistics. http://www.internetworldstats.com/stats.htm. Accessed 20 Aug 2018
2. Chen, C., Schedl, M., Zamani, H.: RecSys Challenge 2018. https://recsys.acm.org/recsys18/challenge/. Accessed 20 Aug 2018
3. Bennet, J., Lanning, S.: The Netflix Prize. https://www.netflixprize.com. Accessed 20 Aug 2018
4. Hu, Y., Koren, Y., Volinsky, C.: Collaborative filtering for implicit feedback datasets. In: Eighth IEEE International Conference on Data Mining, Pisa, Italy, pp. 263–272 (2008)
5. Koren, Y., Bell, R., Volinsky, C.: Matrix factorization techniques for recommender systems. Computer **8**, 30–37 (2009)
6. Koren, Y.: Factorization meets the neighborhood: a multifaceted collaborative filtering model. In: Proceedings of the 14th ACM SIGKDD International Conference on Knowledge Discovery and Data Mining, New York, USA, pp. 426–434, ACM (2008)
7. Rendle, S.: Factorization machines. In: 2010 IEEE 10th International Conference on Data Mining (ICDM), Sydney, Australia, pp. 995–1000. IEEE (2010)
8. Wang, H., Wang, N., Yeung, D.Y.: Collaborative deep learning for recommender systems. In: Proceedings of the 21th ACM SIGKDD International Conference on Knowledge Discovery and Data Mining, New York, USA, pp. 1235–1244, ACM (2015)
9. He, X., Liao, L., Zhang, H., Nie, L., Hu, X., Chua, T.S.: Neural collaborative filtering. In: Proceedings of the 26th International Conference on World Wide Web, Perth, Australia, pp. 173–182. International World Wide Web Conferences Steering Committee (2017)
10. Salakhutdinov, R., Mnih, A., Hinton, G.: Restricted Boltzmann machines for collaborative filtering. In: Proceedings of the 24th International Conference on Machine Learning, New York, USA, pp. 791–798. ACM (2007)

11. Van den Oord, A., Dieleman, S., Schrauwen, B.: Deep content-based music recommendation. In: Advances in Neural Information Processing Systems, pp. 2643–2651 (2013)
12. Kim, D., Park, C., Oh, J., Lee, S., Yu, H.: Convolutional matrix factorization for document context-aware recommendation. In: Proceedings of the 10th ACM Conference on Recommender Systems, New York, USA, pp. 233–240. ACM (2016)
13. He, X., Zhang, H., Kan, M.Y., Chua, T.S.: Fast matrix factorization for online recommendation with implicit feedback. In: Proceedings of the 39th International ACM SIGIR Conference on Research and Development in Information Retrieval, New York, NY, USA, pp. 549–558. ACM (2016)
14. Rendle, S., Freudenthaler, C., Gantner, Z., Schmidt-Thieme, L.: BPR: Bayesian personalized ranking from implicit feedback. In: Proceedings of the Twenty-Fifth Conference on Uncertainty in Artificial Intelligence, Arlington, Virginia, USA, pp. 452–461. AUAI Press (2009)
15. Socher, R., Chen, D., Manning, C., Ng, A.: Reasoning with neural tensor networks for knowledge base completion. In: NIPS, pp. 926–934 (2013)
16. Herlocker, J.L., Konstan, J.A., Borchers, A., Riedl, J.: An algorithmic framework for performing collaborative filtering. In: ACM SIGIR Forum, New York, USA, pp. 227–234. ACM (2017)
17. Sarwar, B., Karypis, G., Konstan, J., Reidl, J.: Item-based collaborative filtering recommendation algorithms. In: Proceedings of the Tenth International Conference on World Wide Web, WWW 2001, New York, USA, pp. 285–295 (2001)
18. Kim, Y.: Convolutional neural networks for sentence classification. In: Proceedings of the 2014 Conference on Empirical Methods in Natural Language Processing (EMNLP), Doha, Qatar, pp. 1746–1751 (2014)
19. Glorot, X., Bordes, A., Bengio, Y.: Deep sparse rectifier neural networks. In: Proceedings of the Fourteenth International Conference on Artificial Intelligence and Statistics, Florida, USA, pp. 315–323 (2011)
20. Kiefer, J., Wolfowitz, J.: Stochastic estimation of the maximum of a regression function. Ann. Math. Stat. **23**(3), 462–466 (1952)
21. MovieLens Dataset. https://grouplens.org/datasets/movielens/. Accessed 20 Aug 2018
22. Bayer, I., He, X., Kanagal, B., Rendle, S.: A generic coordinate descent framework for learning from implicit feedback. In: WWW, Perth, Australia, pp. 1341–1350. International World Wide Web Conferences Steering Committee (2017)
23. Wu, Y., DuBois, C., Zheng, A.X., Ester, M.: Collaborative denoising auto-encoders for top-n recommender systems. In: Proceedings of the Ninth ACM International Conference on Web Search and Data Mining, San Francisco, California, USA, pp. 153–162. ACM (2016)
24. Kingma, D.P., Ba, J.: Adam: a method for stochastic optimization. In: International Conference on Learning Representations (ICLR), San Diego, USA, pp. 1–15 (2014)
25. Liu, Y., Wang, S., Khan, M.S., He, J.: A novel deep hybrid recommender system based on auto-encoder with neural collaborative filtering. Big Data Min. Anal. **1**(3), 211–221 (2018)
26. LeCun, Y., Bottou, L., Bengio, Y., Haffner, P.: Gradient-based learning applied to document recognition. Proc. IEEE **86**(11), 2278–2324 (1998)

A Pedestrian Dead Reckoning System with Error Correction by Landmark Recognition

Khanh Nguyen-Huu[1] and Seon-Woo Lee[1,2(✉)]

[1] Research Institute of Information and Electronic Engineering,
Hallym University, Hallimdaehak-gil,
Chuncheon, Gangwon-do, Republic of Korea
khanhnguyenhuu177@gmail.com, senu@hallym.ac.kr
[2] School of Software, Hallym University,
Hallimdaehak-gil, Chuncheon, Gangwon-do, Republic of Korea

Abstract. Due to the high demand of location-based services in buildings, various indoor positioning methods have been proposed. Among them, the Pedestrian Dead Reckoning (PDR) systems have received much attention because of the widespread deployment of smartphones and no requirement of infrastructure. In this paper, we propose a PDR system to track a pedestrian holding a smartphone while he/she walks in a building. To eliminate the inevitable accumulated error over time, we suggest a method of using location related information from environments. We named it 'landmark', which is defined as a specific area (or point) where the pedestrian passes by in a building such as a stair, a door, a corner or a crossroad. If the system detects a landmark during the walking period, then it will correct the position of the user with the pre-defined position information of the landmark. In this work, we propose one kind of landmark called the door landmark. We have developed a method to detect the landmarks only using the sensors embedded in smartphones. The door landmark can be detected by combining the features from the magnetic sensor and user's walking behavior. Via a set of experiments, we have evaluated the performances of the landmark detection and the position tracking of the system.

Keywords: PDR · Landmark · Door detection · Smartphone · Indoor positioning

1 Introduction

In recent years, the attention and demand of indoor location-based services have increased dramatically, especially the accurate real-time indoor localization due to its broad applications such as emergency rescue or surveillance in the building. The global positioning system (GPS) runs well for outdoor environments, but in case of indoor ones, it cannot reach the good results due to the non-line-of-sight signals and the building's material such as thick, solid materials (e.g. brick, metal, or wood). For the past 20 years, there are many indoor localization systems have been developed to track the user's position accurately. Two main branches that receive the concentrations are

© Springer Nature Singapore Pte Ltd. 2019
J.-H. Kim et al. (Eds.): RiTA 2018, CCIS 1015, pp. 132–143, 2019.
https://doi.org/10.1007/978-981-13-7780-8_11

wireless radio technique and dead reckoning technique. The former one, which involves different types of wireless facility such as WiFi [1], Bluetooth [2], radio frequency identification (RFID) [3], or FM radio [4], can solve the localization problem by two ways: triangulation and fingerprinting. The main limitation of this technique is the high setup cost and time-consuming works [5]. The dead reckoning method, which uses the inertial sensor to estimate the displacement of the user based on the previous position, is also a potential solution for indoor localization. The pedestrian dead reckoning (PDR) (in this work, our system is a Step-and-Heading System (SHS) specifically) is a popular technique due to its lightweight and inexpensive sensors which can be found easily on daily portable devices such as smartphones, smartwatches, or tablets. In PDR systems, the position could be computed by adding the incremental movement representing one walking step, so the error during the estimation of the step length and the heading measurement could result in a big position error after a long time. Especially, the heading error has tremendous effect on the estimation of user's position.

Minimizing the accumulated position error is a challenging topic. There have been various solutions to reduce the error of PDR system. Some researchers applied different kinds of filters such as Kalman filter [6, 7] or particle filter [8, 9] but they require huge computing power for real-time operation. Others have introduced the special locations called landmarks (e.g. stairs, door, corner, etc.) which often exist indoors to reset the position. The reference point of the WiFi fingerprint technique is one example of the landmark. If a system can detect that a user is at a landmark, then it can re-set the user's position with the given location information. In [10], a system has been suggested that compensates for the PDR position with the help of WiFi landmarks to cover the whole walking path. A method also has been developed to match a special area in a map and the recognized user's motion for finding a position [11, 12]. The authors also suggested a method to use structural landmarks in buildings such as using escalator, elevator, and stairs.

Using the magnetic field of the earth could be a promising solution to detect the landmark for correcting the position error since it does not require any extra infrastructure and as a fingerprint, its stability is better than the RSSI of the WiFi. Moreover, almost smart devices include the geomagnetic sensor. In [13, 14], a magnetic-based map matching method which utilizes the stability of magnetic fingerprint, has been introduced to improve the accuracy of the conventional PDR system. Recently, Li et al. [15] have suggested a method to improve the performance of their indoor positioning system by using the extended Kalman filter based on the information from inertial and magnetic sensors, WiFi signals. He and Shin [16] summarized the potential of using geomagnetism in the field of indoor localization, especially the smartphone-based techniques.

This paper presents an indoor navigation system using a smartphone based on PDR and door landmark detection. The proposed method attempts to estimate the position of a user by using the improved PDR method with error correction when it detects door landmarks. The improved PDR method does not require prior knowledge (or restrictions) on holding a smartphone, unlike other methods. Based on the recognized holding style, the system can estimate the position more accurately. The proposed door landmark recognition method uses the dynamic time warping (DTW) algorithm based on the magnetic sensor output signals. Moreover, we analyzed the characteristics of magnetic signals when a user passes a door. In this work, we only consider a wood type door even there are many types of a door like metal, glass, and so on. We could make a fingerprint for one door with four templates due to four states such as: passing through the opened door, passing through the closed door with opening and closing phases, and two moving directions (in/out). If the system detects the door landmark (e.g. the user passes the detected door), then the position coordinator re-sets the current position with the given location of the door.

The rest of the paper is organized as follows. Section 2 describes the proposed system in detail. The performance evaluations and analyses are presented in Sect. 3. Finally, Sect. 4 concludes the paper and mentions the future work.

2   Proposed System

2.1   Overview

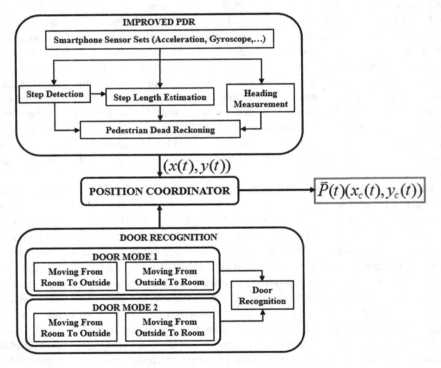

Fig. 1. Block diagram of the proposed system.

The block diagram of the proposed system is presented in Fig. 1, which composes of two main components: the improved PDR and the door recognition components.

The improved PDR component is a typical SHS based on recognition of holding styles that has been proposed in our previous work [17]. In order to free the user from a fixed holding style, the four smartphone holding style is classified firstly as holding the phone on hand in front of the body, holding the phone parallel with ears while making a phone call, swinging the phone during walking phase, and putting the phone into the front pants' pockets. With a set of features extracted from different sensors such as accelerometer, gyroscope, etc. the decision tree method is applied. For each detected holding style, while the user is walking, the improved PDR component attempts to detect the walking step, then estimates the step length and measures heading orientation. The holding style will affect the way of treating each factor of PDR. The output of the PDR component is the 2D coordinate resulted as $(x(t), y(t))$.

The door landmark recognition component tries to detect that a user passes through a door. The proposed method uses the DTW method with the fingerprint based on the sequences of magnetic signals. As shown in Fig. 1, there are two modes for a door to represent the current state of that door (open/close). Therefore, we could get different templates depending on the door state and user moving direction. By using the magnetic values extracted from the smartphone, the system attempts to figure out the door crossing behavior while the user is walking, then corrects the position of the user if some conditions are met. The position of the pedestrian after calibration is presented as $(x_c(t), y_c(t))$.

The position coordinator plays a role as the center to decide to update the pedestrian position. For every new walking step detected, based on the landmark information and satisfied conditions, the location of the pedestrian is compensated.

In this work, we assumed that a user holds a smartphone with only one holding style (holding the phone on hand in front of the body, HA) because the magnetic signals are totally different depending on the orientation of the magnetic sensor (i.e. holding styles). Therefore, for the convenience of implementation, we fixed the holding style. We expect to expand the numbers of templates for a fingerprint with respect to the holding styles.

2.2   The Improved PDR Component

The sampling rate for reading the set of sensors is 30 Hz. The scheme firstly classifies the smartphone holding styles. Four holding styles are considered: holding the phone on hand in front of the body (HA), holding the phone parallel with ears while making a phone call (CA), swinging the phone during walking (SW), and putting the phone into the front pants' pocket (PO). A decision tree method (J48 classifier) is used for classification due to its ease of implementing and fast predictions. The time domain features are used from the accelerometer and the gyroscope: mean, standard deviation, variance, maximum, and minimum. The total number of features is 30. Each feature was computed using the sliding window approach with the overlap of 60% of the data. For each holding style, the classifier was trained with the data of 10 min.

After the holding style is detected, the step detection method performs. Different sets of sensor inputs (i.e. accelerometer for HA and CA and gyroscope for SW and PO) are used to detect a walking step. The proposed method tries to determine a new step by finding two peaks of the different sensor inputs (acceleration or angular velocity). The step length also is estimated based on the speed of walking and the strength of sensors values. The mean error of detecting the walking step is 4.45% and for the step length estimation, it is 4.48%, details are written in our previous work [18].

In step-and-heading systems, accurate measurement of the heading of the pedestrian is an essential factor for the performance of tracking. In our previous work [19], the four main holding styles and 34 sub-cases of them were recognized and the scheme could calibrate the relative heading offset based on the detected holding style. In addition, a complementary filtering technique was used to compute the heading of user's walking from the signal of a gyroscope with the accelerometer and the magnetometer. Then some heuristic methods such as quantization (i.e. the 8 azimuth directions only are used for updating the position) method with hysteresis, and the alignment of missing true North direction (i.e. 15° in case of our building) were applied to estimate the final heading direction. The proposed method showed the mean error of 20° during the walking of a user while changing the holding styles.

2.3 Door Recognition Method

2.3.1 Door Crossing Behavior

When a user passes through a door, there could be two cases: the door is already opened (door mode 1) and the door is closed (door mode 2). In case of the door mode 1, the user just walks through the door to enter without any stopping moments. However, in door mode 2, the user should stop in front of the door, pushes/pulls the door, walks again, and then closes the door. Thus, the passing time of this mode is longer than the door mode 1.

Inspired by the idea of Zhao et al. [20], in this paper, we also focus on using magnetic field values to detect the door crossing behavior of the user. At first, we observe the magnitudes of the vectors of the magnetic sensor signal, $|\vec{m}|$, $\vec{m} = (m_x, m_y, m_z)$ when a pedestrian crosses the doors. There are four doors in this experiment and for each door, we observed the result in two door modes. Figures 2 and 3 show the changing of $|\vec{m}|$ when a pedestrian walks through the same door in two door modes. The lower graphs (True Value legend) represent the real duration between the starting and ending of door passing period. Number 1 and 3 show the cases of going out of the room, and number 2 and 4 are the cases of going into the room. The door material also could affect much the magnetic field signals as well known. In this work, we only consider the wood type door. As shown in two figures, when the user walks through the door, whatever the door mode (1 or 2) or the direction (moving out or moving into), there always exists a "V" pattern of $|\vec{m}|$, thus this can be used as one clue to prove the door crossing behavior. However, there also exist many similar patterns with the above one during the walking phase of the pedestrian. Therefore, using only the changing of $|\vec{m}|$ is not enough for detecting the door crossing behavior.

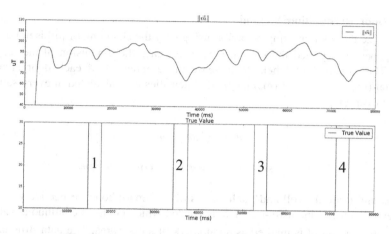

Fig. 2. A pedestrian walks through an opened door.

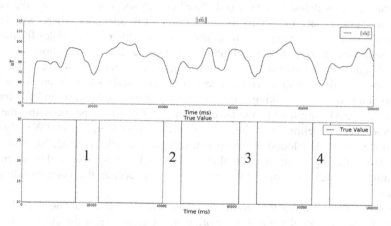

Fig. 3. A pedestrian walks through a closed door.

We also did the same experiment with different door material such as metallic and glass doors, and the "V" patterns occur the clearest in case of metallic doors, than the case of wooden doors, and the worst result for glass doors. Therefore, it could be said that the door material can affect to the door recognition result, which also was pointed out in [20]. Moreover, we could see the "V" patterns of $|\vec{m}|$ even when the pedestrian walks at different walking speeds (i.e. slow, normal, and fast).

At last, when the pedestrian changes his smartphone holding styles from holding it in front of his body (HA-default) to holding the phone parallel with ears while making a phone call (CA), swinging the phone during walking phase (SW), or putting the phone into the front pants' pockets (PO), the CA holding style can show the "V" pattern but it does not occur when the pedestrian keeps the phone in two other styles (SW and PO). Therefore, the holding style also affects to the door recognition result.

2.3.2   Door Recognition Method

To recognize the door crossing behavior, using only the changing of $|\vec{m}|$ is not enough. In this paper, we applied the idea in [15] to increase the dimension of the magnetic values. Their method will help to increase the uniqueness of each door area. The horizontal (m_h) and vertical (m_v) magnetic intensities are calculated at every sampling time as in Eq. (1)

$$m_h = \sqrt{|\vec{m}| - m_v} \tag{1}$$

$$m_v = -\sin\theta.m_x + \sin\varphi.m_y + \cos\theta\cos\varphi.m_z \tag{2}$$

where φ and θ are the roll and pitch values of the smartphone, respectively.

Then a magnetic-intensity-template is created which contains continuous values of $|\vec{m}|$, m_h, and m_v and it is marked as a landmark of a door area. The data structure of a door landmark is formatted as {ID, mode, direction, template, position (x_{LM}, y_{LM})}, where the mode is defined as "opened door" or "closed door", and the direction is defined as "going out" or "going in". Therefore, for one door, there will be four fingerprints having the same position information with different templates for magnetic signal sequences to cover four combinations of two modes and directions.

The lengths of the measured magnetic signal sequence are different due to the different passing time. In this work, the length of door mode 1 is 150 samples which is corresponding to 5 s of walking duration (sampling rate is 30 Hz), and for the door mode 2, it is 300 samples (10 s). To compare the difference between the measured signal and the pre-defined template, the dynamic time warping (DTW) algorithm, which is originally developed in the speech recognition field, is used. By using DTW method the matching scores are calculated and saved to a list. Next, the system finds the minimum matching score, thus the system can decide the corresponding door landmark.

However, there can be the landmark mismatches occurred if one door landmark accidentally has the smaller matching score than the right one. If the current position is updated to the landmark position at this time, there may cause a big location error between the true position and the compensated position. In order to avoid this, we consider the geometric distance (g_d) on 2D map between the positions of the user and the landmark. It is assumed that the shapes of the user and the landmark area are the circle shapes with the radius of each one is 1.5 m and 5 m. If the geometric distance g_d is less than 6.5 m (i.e. collision case) could also be used another condition to compensate the current position by detecting a door.

Moreover, we also introduce a condition that the interval between consecutive door detections should be bigger than a given number of continuous walking steps (continuous_step). This condition could solve the re-correction problem when the user walks slowly so he/she cannot walk out of the area where the matching score of the detected door still is the smallest one. The number of continuous walking step is determined empirically as 15 steps.

3 Experimental Results

3.1 Experiment Setup

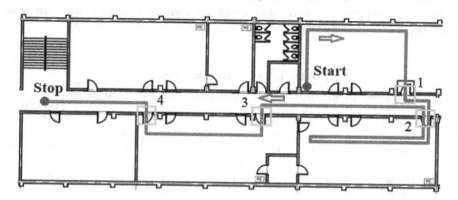

Fig. 4. The scenario with landmarks. (Color figure online)

To evaluate the performance of the proposed system, several experiments were conducted by two users (all males). Two different smartphones (LG V35 and Galaxy Note 4) were used to perform the experiments. All the experiments were executed in the Engineering building, Hallym University. Figure 4 shows the moving path of a user which contains four door landmarks. The yellow rectangles are the door landmarks. The red dots are the start and stop positions. The total length of the walking path shown as green line is 110 m. Two kinds of experiments were carried out. The first one is to validate the performance of the door recognition and the next one is to evaluate the positioning performance of the proposed system.

3.2 Door Detection Performance

In this experiment, two users walked on the given path. There were four cases for each door: opened door and going out or into the room, closed door and going out or into the room. Each case was conducted 5 times by each user, which means 20 times for one door landmark, so totally 160 times for two users on four doors. Table 1 shows the performance of recognizing the door landmark with the confusion matrix.

As shown in Table 1, the door's name is represented as D, and two modes are represented as M1 (opened door) and M2 (closed door), respectively. The performance of the door recognition is quite good with the highest error numbers are 3 times over 20 times (D1-M1) conducting the experiment for one door. The main reason for this is due to the close distance between D1 and D2, which is approximate 2.5 m, thus, the door areas (in circles) covered by D1 and D2 can be overlapped. Therefore, if the pedestrian walks into D2 but the matching score of D1 is smaller than D2, then the system can accidentally recognize a wrong result. For the case of D3 and D4, the distance between these doors is bigger than 5 m, therefore the door areas covered by these doors are not overlapped and not coming to the phenomenon of D1 and D2. The lowest and highest error rates were 0% for D3 and D4 and 15% for D1-M1. Among the total of 160 times conducting the experiment for four doors, there are 7 errors, thus the average error rate was 4.38%.

Table 1. Door recognition results.

	Recognized	D1-M1	D1-M2	D2-M1	D2-M2	D3-M1	D3-M2	D4-M1	D4-M2	Error rate (%)
Landmark	D1-M1	17	0	3	0	0	0	0	0	15
	D1-M2	0	18	0	2	0	0	0	0	10
	D2-M1	1	0	18	1	0	0	0	0	10
	D2-M2	0	2	0	18	0	0	0	0	10
	D3-M1	0	0	0	0	20	0	0	0	0
	D3-M2	0	0	0	0	0	20	0	0	0
	D4-M1	0	0	0	0	0	0	20	0	0
	D4-M2	0	0	0	0	0	0	0	20	0

3.3 Tracking Performance

This experiment proves the full system performance. The two users walked on the trajectory described in Fig. 4. Each user walked three times in the trajectories, which means 6 times for two users with the length of 660 m. During the walking phase, the users were asked to keep the smartphone in front of his body. The proposed system performance will be compared with the improved PDR, which mean each user walk three times for each case.

In Fig. 5, the blue dots represent the walking path of the user using the proposed system and the yellow stars represent the walking path using the PDR. The red stars are the landmark positions where the location compensation occurs (i.e. the landmarks that satisfied the necessary conditions). The green pentagons are the previous positions of the pedestrian before location compensating. The figure shows that the walking shape of the user using the proposed system is more similar to the true walking path compared with the walking path using PDR.

With the correction of the selected landmark positions, the proposed system achieved a better result than the improved PDR. Figure 6 shows the accumulated error results of the two methods. At the 50th percentile, the average positioning error of the proposed method for two pedestrians is within 3.78 m and at the 80th percentile, it is within 5.80 m. Meanwhile, at 50th and 80th percentile, the improved PDR got the average error of 5.26 m and 8.20 m. Therefore, the positioning error of the proposed method is decreased by 29.27% compared to the PDR only.

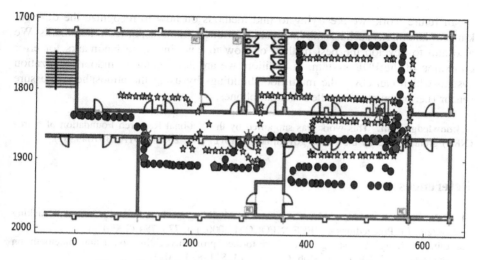

Fig. 5. Positioning result for user 2. (Color figure online)

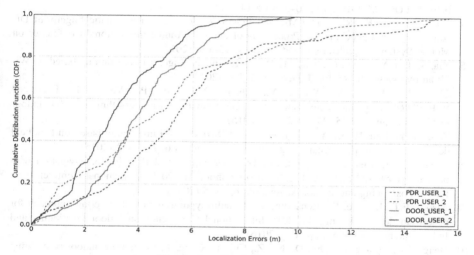

Fig. 6. Localization error CDF.

4 Conclusion

This paper presented an indoor localization system which is implemented on a smartphone. A PDR-based system is introduced with the aid of door landmarks recognition to correct the accumulated error of the conventional PDR. The recognition of door landmark uses the DTW method with the magnetic signals. The door recognition accuracy can reach to 95.62%. The system showed a better performance of tracking compared to the PDR method only, with 29.27% error reduction.

In future work, we are trying to find methods on how to recognize the different kinds of landmarks such as stair landmarks, turning landmarks, or radio landmarks. We also aim to develop the best method on how to combine these landmarks for error correction of the PDR system. In addition, we are developing an indoor localization system which can cover the multi-floor buildings by using the atmospheric pressure sensor (i.e. barometer) installed in smartphones.

Acknowledgement. This work was supported by the National Research Foundation of Korea (NRF) grant funded by the Korea government (MSIT) (No. 2018R1D1A3B07049887).

References

1. Bahl, P., Padmanabhan, V.N.: RADAR: an in-building RF-based user location and tracking system. In: Proceedings of IEEE INFOCOM 2000, pp. 775–784 (2000)
2. Liu, S., Jiang, Y., Striegel, A.: Face-to-face proximity estimation using bluetooth on smartphones. IEEE Trans. Mob. Comput. **13**, 811–823 (2014)
3. Huang, W., Xiong, Y., Li, X.Y., Lin, H., Mao, X., Yang, P., et al.: Shake and walk: acoustic direction finding and fine-grained indoor localization using smartphones. In: IEEE INFOCOM 2014, pp. 370–378 (2014)
4. Yoon, S., Lee, K., Rhee, I.: FM-based indoor localization via automatic fingerprint DB construction and matching. In: Proceeding of the 11th Annual International Conference on Mobile Systems, Applications, and Services, Taiwan (2013)
5. Xia, S., Liu, Y., Yuan, G., Zhu, M., Wang, Z.: Indoor fingerprint positioning based on Wi-Fi: an overview. ISPRS Int. J. Geo-Inf. **6**, 135 (2017)
6. Chen, G., Meng, X., Wang, Y., Zhang, Y., Tian, P., Yang, H.: Integrated WiFi/PDR/Smartphone using an unscented Kalman filter algorithm for 3D indoor localization. Sensors **15**, 24595–24614 (2015)
7. Zhuang, Y., Lan, H., Li, Y., El-Sheimy, N.: PDR/INS/WiFi integration based on handheld devices for indoor pedestrian navigation. Micromachines **6**, 793–812 (2015)
8. Ascher, C., Kessler, C., Wankerl, M., Trommer, G.F.: Dual IMU indoor navigation with particle filter based map-matching on a smartphone. In: 2010 International Conference on Indoor Positioning and Indoor Navigation, pp. 1–5 (2010)
9. Alaoui, F.T., Betaille, D., Renaudin, V.: A multi-hypothesis particle filtering approach for pedestrian dead reckoning. In: 2016 International Conference on Indoor Positioning and Indoor Navigation (IPIN), pp. 1–8 (2016)
10. Deng, Z.-A., Wang, G., Qin, D., Na, Z., Cui, Y., Chen, J.: Continuous indoor positioning fusing WiFi, smartphone sensors and landmarks. Sensors **16**, 1427 (2016)
11. Elhoushi, M., Georgy, J., Noureldin, A., Korenberg, M.J.: Motion mode recognition for indoor pedestrian navigation using portable devices. IEEE Trans. Instrum. Meas. **65**, 208–221 (2016)
12. Alaoui, F.T., Renaudin, V., Betaille, D.: Points of interest detection for map-aided PDR in combined outdoor-indoor spaces. In: 2017 International Conference on Indoor Positioning and Indoor Navigation (IPIN), pp. 1–8 (2017)
13. Shu, Y., Bo, C., Shen, G., Zhao, C., Li, L., Zhao, F.: Magicol: indoor localization using pervasive magnetic field and Opportunistic WiFi sensing. IEEE J. Sel. Areas Commun. **33**, 1443–1457 (2015)

14. Xie, H., Gu, T., Tao, X., Ye, H., Lu, J.: A reliability-augmented particle filter for magnetic fingerprinting based indoor localization on smartphone. IEEE Trans. Mob. Comput. **15**, 1877–1892 (2016)
15. Li, Y., Zhuang, Y., Zhang, P., Lan, H., Niu, X., El-Sheimy, N.: An improved inertial/WiFi/magnetic fusion structure for indoor navigation. Inf. Fusion **34**, 101–119 (2017)
16. He, S., Shin, K.G.: Geomagnetism for smartphone-based indoor localization: challenges, advances, and comparisons. ACM Comput. Surv. **50**, 1–37 (2017)
17. Nguyen-Huu, K., Lee, K., Lee, S.: An indoor positioning system using pedestrian dead reckoning with WiFi and map-matching aided. In: 2017 International Conference on Indoor Positioning and Indoor Navigation (IPIN), pp. 1–8 (2017)
18. Nguyen-Huu, K., Seong-Won, S., Je, Y.-Y., Song, C.-G., Lee, S.-W.: A step detection and step length estimation method based on different holding styles estimation. In: KCC 2015, Jeju, pp. 359–361 (2015)
19. Nguyen-Huu, K., Song, C.G., Lee, S.-W., Lee, K.: A heading estimation based on smartphone holding styles. In: HCI Korea, pp. 1–7 (2016)
20. Zhao, Y., Qian, C., Gong, L., Li, Z., Liu, Y.: LMDD: light-weight magnetic-based door detection with your smartphone. In: 44th International Conference on Parallel Processing, pp. 919–928 (2016)

Effective Indoor Robot Localization by Stacked Bidirectional LSTM Using Beacon-Based Range Measurements

Hyungtae Lim and Hyun Myung$^{(\boxtimes)}$

Urban Robotics Lab, KAIST,
291, Daehak-ro, Yuseong-gu, Daejeon, Republic of Korea
{shapelim,hmyung}@kaist.ac.kr
http://urobot.kaist.ac.kr

Abstract. In this paper, we propose a stacked bidirectional Long Short-Term Memory (stacked Bi-LSTM) for accurate localization of a robot. Using deep learning, the proposed structure directly maps range measurements from beacons into robot position. This operation non-linearly maps the relationship not only considering the long-range dependence of sequential distance data but also using the correlation of the backward information and the forward information of the sequence of each time step by virtue of its bidirectional architecture. Our stacked bidirectional LSTM structure exhibits better estimates of robot positions than other RNN structure units on the simulated environment. In addition, experiments suggest that even if the robot position is not included in the training dataset, our method is able to predict robot positions with small errors through sequential distance data.

Keywords: Trilateration · LSTM · Mobile robot ·
Bidirectional LSTM · Recurrent Neural Networks

1 Introduction

Trilateration is a conventional algorithm for locating a vehicle in the metropolitan area by range measurements between the vehicle and fixed beacon sensors [1]. Due to the convenience of trilateration that estimates the position of a receiver of range sensors if one only knows range measurement, trilateration algorithm has been widely incorporated into robotics fields, especially utilized in the indoor environment to estimate the position of an object by distance measurements

H. Lim—This material is based upon work supported by the Ministry of Trade, Industry & Energy (MOTIE, Korea) under Industrial Technology Innovation Program. No. 10067202, 'Development of Disaster Response Robot System for Lifesaving and Supporting Fire Fighters at Complex Disaster Environment'.

© Springer Nature Singapore Pte Ltd. 2019
J.-H. Kim et al. (Eds.): RiTA 2018, CCIS 1015, pp. 144–151, 2019.
https://doi.org/10.1007/978-981-13-7780-8_12

obtained from range sensors such as UWB, ultrasonic, laser-based beacon sensors [2–4]. Specifically, range-only Simultaneous Localization and Mapping (RO-SLAM) methods are utilized popularly, which not only estimate the position of the receiver of range sensors, but also localize the position of range sensors regarded as features on a map, and studies have been conducted continuously in terms of probability-based approach [5–8].

In the meantime, as deep learning age has come [9], various kinds of deep neural architectures have been proposed for many tasks related to robotics field, such as detection [10–12], navigation [13,14], pose estimation [15], and so on. Especially, recurrent neural networks (RNNs), originated from Natural Language Process (NLP) area [16], have been shown to achieve better performance in case of dealing with time variant information, thereby RNNs are widely utilized such as not only speech recognition, but also pose estimation and localization [15,17–20].

In this paper, we propose a deep learning-based localization method by stacked bidirectional Long Short-Term Memory (stacked Bi-LSTM) for more accurate localization of the robot. Using deep learning, our structure directly learns the end-to-end mapping between range measurements and robot position. This operation non-linearly maps the relationship not only considering the long-range dependence of sequential distance data by the LSTM, but also using the correlation of the backward information and the forward information of the sequence of each time step by virtue of its bidirectional architecture (Fig. 1).

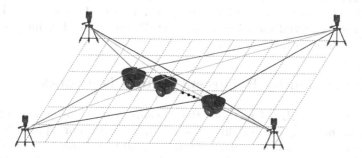

Fig. 1. System overview. A mobile robot localizes its own pose through range measurements and the derivative of range measurements.

2   Related Works

In this section, we briefly survey previous researches closely focused on Long Short-Term Memory (LSTM) model and applications of LSTMs to solve domain problems.

LSTM. LSTM is a type of Recurrent Neural Networks (RNNs) that has loops so that infer output based on not only the input data, but also the internal state formed by previous information. In other words, while the RNN deals with sequential data, the network has remembered the previous state generated by past inputs and might be able to output the present time step via internal state and input, which is very similar to filtering algorithms.

However, RNNs often have a *vanishing gradient problem*, i.e., RNNs fail to propagate the previous matter into present tasks as time step gap grows by. In other words, RNNs are not able to learn to store appropriate internal states and operate on long-term trends. That is the reason why the Long Short-Term Memory (LSTM) architecture was introduced to solve this long-term dependency problem and make the networks possible to learn longer-term contextual understandings [21]. By virtue of the LSTM architecture that has memory gates and units that enable learning of long-term dependencies [22], LSTM are widely used in most of the deep learning research areas and numerous variations of LSTM architectures have been studied.

Applications of LSTMs. There are many variations of LSTM architecture. As studies of deep learning are getting popular, various modified architectures of LSTM have been proposed for many tasks in a wide area of science and engineering. Because LSTM is powerful when dealing with sequential data and inferring output by using previous inputs, LSTM is utilized to estimate pose by being attached to the end part of deep learning architecture [18–20] as a stacked form of LSTM. In addition, LSTM takes many various data as input; LSTM is exploited for sequential modeling using LiDAR scan data [17], images [15,18], IMU [23], a fusion of IMU and images [24].

3   Trilateration by Recurrent Neural Networks

Unlike existing RO-SLAM methods that localize a robot by probability-based approach or filtering method like Kalman filter, etc., our approach localize robot's position by letting the networks be trained by distance data and ground truth of the robot's position. We express the problem statement for the localization of the robot using range measurements. The training input data set is formulated as follows:

$$L = (X_t, Y_t) \tag{1}$$

where $X_t = \{(l_1, l_2, \ldots, l_m)_t, m = 1, \ldots, M\}$ denotes input range data from range sensors and M is the number of UWB sensors at time t. Ground truth of the robot's 2D position is denoted as $Y_t = \{(x_t, y_t)\}$.

Let Θ be the parameters of a RNN model and assume that the trained RNN model could be expressed as conditional probability as follows:

$$P(Y_t|X_t) = p((x_t, y_t)|(l_1, \ldots, l_m)_{t-T+1}, (l_1, \ldots, l_m)_{t-T+2} \ldots, (l_1, \ldots, l_m)_t) \quad (2)$$

where T indicates sequential length of input to LSTM. Then, our final goal is to find optimal parameters Θ^* for localization by minimizing mean square error (MSE) of Euclidean distance between ground truth position Y_k and estimated position $\hat{Y}_k$ as follows:

$$\Theta^* = \underset{\Theta}{\operatorname{argmin}} \frac{1}{N} \sum_{k=1}^{N} \parallel Y_k - \hat{Y}_k \parallel^2 \quad (3)$$

4 Simulations

In this section, we describe the simulation environment and the type of LSTM used for the simulation.

4.1 Training/Test Dataset

To verify our proposal that RNNs can estimate the robot's position through varying range data, we set the experiment on the simulated environment and generate range data set which corresponds to the position with 10% noise error and let RNN be trained using these range data. Train data are just zigzag paths that generate Δx or Δy changes respectively. Test data are two types; a diagonal oval path and an arbitrary path, which are shown as Fig. 3(a) and (d). There's no same path between train data set and test data set. In addition, unlike train data set, x and y directions change at the same time. Thus, we also check the capability of RNNs to estimate the position in the regions not included in the train data set by variation of range data over time as input (Fig. 2).

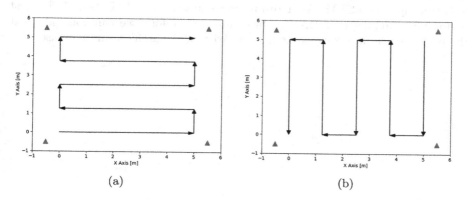

(a)   (b)

Fig. 2. Train dataset consists of (a) the horizontal zigzag path and (b) the vertical zigzag path. And red triangular points indicate positions of 4 UWB sensors: $(-0.5, -0.5)$, $(-0.5, 5.5)$, $(5.5, -0.5)$, $(5.5, 5.5)$. (Color figure online)

4.2 RNN Architectures

Various kinds of RNN models are tested for robot localization through range measurements. We trained 4 kinds of RNNs; LSTM [21], GRU [25], Bi-LSTM [26], stacked Bi-LSTM, which are basic architecture units in many RNNs researches. Using deep learning, these structures directly learn the end-to-end mapping between range data and robot position.

Table 1. RMSE of each RNN model for the test data.

	RMSE of localization [cm]			
	LSTM	GRU	Bi-LSTM	Stacked Bi-LSTM
Test1	9.6839	15.5183	7.5110	**6.3788**
Test2	10.3990	18.1394	7.6592	**6.6906**

5 Results

The results of trajectory prediction are shown in Fig. 3(a) and (d) and Root-Mean-Squared Error (RMSE) are shown in Table 1. Performance is better in order of stacked Bi-LSTM, Bi-LSTM, LSTM and GRU. In case of GRU, it has only two gates which is less complex structure than LSTM [27]. However, due to GRU's less complexity, GRU has less number of neurons than LSTM so their non-linear mapping achieves less performance. Likewise, Bi-LSTM consists of two LSTMs to process sequence in two directions so that infer output using the correlation of the backward information and the forward information of the sequences of each time step with its two separate hidden layers. Thus, Bi-LSTM has better nonlinear mapping capability than LSTM. For similar reasons, stacked Bi-LSTM is the architecture that stacks two Bi-LSTMs, so inference performance is better than Bi-LSTM. As a result, the stacked Bi-LSTM showed the best performance among unit RNN architectures. Therefore, we can conclude that the performance improves as the non-linearity of the architecture increases.

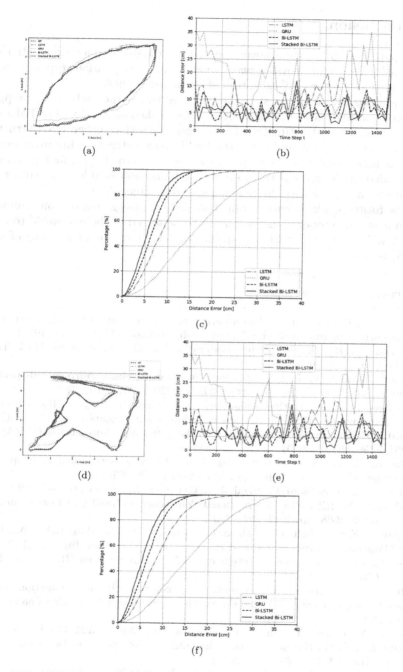

Fig. 3. The results of trajectories of each test path: (a) the curved oval path and (d) the arbitrary path. (b) the distance error with time step (*the lower is better*) (c) CDF (*The trial that reaches 100% faster is better*) of the curved diagonal path in (a). (e) the distance error with time step (f) CDF of the arbitrary path in (d)

6 Conclusion

In this paper, we proposed a novel approach to range-only measurements localization using recurrent neural network models and tested various types of LSTM models for more accurate localization of the mobile robot.

Using deep learning, our structure directly learns the end-to-end mapping between distance data and robot position. The stacked bidirectional LSTM structure exhibits the best estimates of robot positions than other RNN structure units. Therefore, we conclude that the LSTM-based structure improves performance as non-linearity of structures increase and even if the robot position is not included in the ground truth dataset, our method is able to predict robot positions with small errors through sequential distance data.

As a future work, because train/test dataset are generated on simulated environment, the proposed method needs to be tested in the real-world to check whether RNNs can deal with multipath problems and line of non-line of sight (NLOS) issues.

References

1. Staras, H., Honickman, S.: The accuracy of vehicle location by trilateration in a dense urban environment. IEEE Trans. Veh. Technol. **21**(1), 38–43 (1972)
2. Thomas, F., Ros, L.: Revisiting trilateration for robot localization. IEEE Trans. Robot. **21**(1), 93–101 (2005)
3. Cho, H., Kim, S.W.: Mobile robot localization using biased chirp-spread-spectrum ranging. IEEE Trans. Ind. Electron. **57**(8), 2826–2835 (2010)
4. Raghavan, A.N., Ananthapadmanaban, H., Sivamurugan, M.S., Ravindran, B.: Accurate mobile robot localization in indoor environments using bluetooth. In: 2010 IEEE International Conference on Robotics and Automation (ICRA), pp. 4391–4396. IEEE (2010)
5. Blanco, J.L., González, J., Fernández-Madrigal, J.A.: A pure probabilistic approach to range-only slam. In: ICRA, pp. 1436–1441 (2008). Citeseer
6. Blanco, J.L., Fernández-Madrigal, J.A., González, J.: Efficient probabilistic range-only slam. In: IEEE/RSJ International Conference on Intelligent Robots and Systems, IROS 2008, pp. 1017–1022. IEEE (2008)
7. Fabresse, F.R., Caballero, F., Maza, I., Ollero, A.: Undelayed 3D RO-SLAM based on Gaussian-mixture and reduced spherical parametrization. In: 2013 IEEE/RSJ International Conference on Intelligent Robots and Systems (IROS), pp. 1555–1561. Citeseer (2013)
8. Shetty, N.S.: Particle filter approach to overcome multipath propagation error in slam indoor applications. Ph.D. thesis, The University of North Carolina at Charlotte (2018)
9. LeCun, Y., Bengio, Y., Hinton, G.: Deep learning. Nature **521**(7553), 436 (2015)
10. Lenz, I., Lee, H., Saxena, A.: Deep learning for detecting robotic grasps. Int. J. Robot. Res. **34**(4–5), 705–724 (2015)
11. Cai, Z., Fan, Q., Feris, R.S., Vasconcelos, N.: A unified multi-scale deep convolutional neural network for fast object detection. In: Leibe, B., Matas, J., Sebe, N., Welling, M. (eds.) ECCV 2016. LNCS, vol. 9908, pp. 354–370. Springer, Cham (2016). https://doi.org/10.1007/978-3-319-46493-0_22

12. Smith, H.H.: Object detection and distance estimation using deep learning algorithms for autonomous robotic navigation (2018)
13. Zhu, Y., et al.: Target-driven visual navigation in indoor scenes using deep reinforcement learning. In: 2017 IEEE International Conference on Robotics and Automation (ICRA), pp. 3357–3364. IEEE (2017)
14. Hamandi, M., D'Arcy, M., Fazli, P.: DeepMoTIon: learning to navigate like humans. arXiv preprint arXiv:1803.03719 (2018)
15. Walch, F., Hazirbas, C., Leal-Taixe, L., Sattler, T., Hilsenbeck, S., Cremers, D.: Image-based localization using LSTMs for structured feature correlation. In: International Conference on Computer Vision (ICCV), pp. 627-637 (2017)
16. Elman, J.L.: Finding structure in time. Cogn. Sci. **14**(2), 179–211 (1990)
17. Gladh, S., Danelljan, M., Khan, F.S., Felsberg, M.: Deep motion features for visual tracking. In: 2016 23rd International Conference on Pattern Recognition (ICPR), pp. 1243–1248. IEEE (2016)
18. Wang, S., Clark, R., Wen, H., Trigoni, N.: DeepVo: towards end-to-end visual odometry with deep recurrent convolutional neural networks. In: 2017 IEEE International Conference on Robotics and Automation (ICRA), pp. 2043–2050. IEEE (2017)
19. Kendall, A., Grimes, M., Cipolla, R.: PoseNet: a convolutional network for real-time 6-DOF camera relocalization. In: Proceedings of the IEEE International Conference on Computer Vision, pp. 2938–2946 (2015)
20. Turan, M., Almalioglu, Y., Araujo, H., Konukoglu, E., Sitti, M.: Deep EndoVo: a recurrent convolutional neural network (RCNN) based visual odometry approach for endoscopic capsule robots. Neurocomputing **275**, 1861–1870 (2018)
21. Hochreiter, S., Schmidhuber, J.: Long short-term memory. Neural Comput. **9**(8), 1735–1780 (1997)
22. Zaremba, W., Sutskever, I.: Learning to execute. arXiv preprint arXiv:1410.4615 (2014)
23. Ordóñez, F.J., Roggen, D.: Deep convolutional and LSTM recurrent neural networks for multimodal wearable activity recognition. Sensors **16**(1), 115 (2016)
24. Clark, R., Wang, S., Wen, H., Markham, A., Trigoni, N.: VINet: visual-inertial odometry as a sequence-to-sequence learning problem. In: AAAI, pp. 3995–4001 (2017)
25. Cho, K., et al.: Learning phrase representations using RNN encoder-decoder for statistical machine translation. arXiv preprint arXiv:1406.1078 (2014)
26. Schuster, M., Paliwal, K.K.: Bidirectional recurrent neural networks. IEEE Trans. Sig. Process. **45**(11), 2673–2681 (1997)
27. Chung, J., Gulcehre, C., Cho, K., Bengio, Y.: Empirical evaluation of gated recurrent neural networks on sequence modeling. arXiv preprint arXiv:1412.3555 (2014)

Track Management for Distributed Multi-target Tracking in Sensor Network

Woo-Cheol Lee and Han-Lim Choi[✉]

Korea Advanced Institute of Science and Technology,
Daejeon 34141, Korea
wclee@lics.kaist.ac.kr, hanlimc@kaist.ac.kr

Abstract. This paper addresses track management for multiple target tracking (MTT) with a sensor network. Track management is needed for track generation and extinction when the targets set is unknown. Based on a consensus-based fusion algorithm, we develope a MTT algorithm that includes the measurement-to-track association (M2TA) and track management. It can be effectively applied even when the sensor detection range is limited and the field-of-view (FOV)s of each sensor are different. Numerical examples are presented in a multi-sensor multi-target scenario to verify that the proposed algorithm works properly in various network structures and clutter environments.

Keywords: Track management · Multiple target tracking · Sensor network · Sensor Fusion

1 Introduction

The target tracking problem using the sensor network is important in various applications such as surveillance and reconnaissance missions using multiple agents. Many researches have been conducted from centralized fusion to fully decentralized fusion for target tracking in sensor networks.

This paper presents a track management method for fully distributed multiple target tracking algorithm. [1] proposed a Kalman Consensus Filter (KCF) for distributed target tracking in sensor networks. This filter cooperatively derives consensus on target state information through exchange of target information between sensor nodes. And has been extensively used in the field of cooperative target state estimation and control [2–5].

Since KCF is an algorithm for a single target tracking, research has also been conducted to deal with multiple targets. [6] presented a modified KCF algorithm for multiple target tracking using M2TA (Measurement to Track Association) and T2TA (Track to Track Association) method. But it is difficult to apply this method directly if the target set is unknown.

© Springer Nature Singapore Pte Ltd. 2019
J.-H. Kim et al. (Eds.): RiTA 2018, CCIS 1015, pp. 152–163, 2019.
https://doi.org/10.1007/978-981-13-7780-8_13

This paper presents a GNN [7] (Global Nearest Neighborhood)-based KCF algorithm that includes track management for multiple target tracking. By including track management, it is applicable to the MTT problems whose target set is unknown, the sensor range is limited and spatially distributed in different locations. To do this, we perform data association on local measurements and data arriving from neighboring nodes, and manage the tracks using association results. For track management, four track states (undefined, initialization, track, extinction) were defined and a logic in which the track state transition occurred according to the M2TA result was constructed. The overall algorithm configuration is shown in Fig. 1.

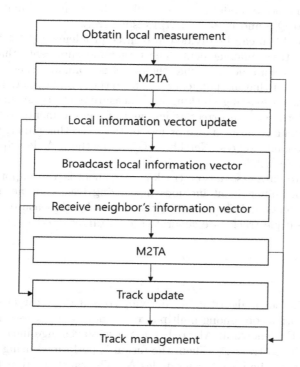

Fig. 1. Flow diagram of GNN-KCF with track management

The rest of the paper is organized as follows. Section 2 describes backgrounds including MTT problem, KCF and GNN algorithm. Section 3 describes track management method we used. Section 4 presents the algorithm developed in this paper. In Sect. 5, numerical examples are presented. Section 6 provides conclusion.

2   Background

2.1   MTT Problem

In the MTT problem, the target model and the sensor model are expressed as follows.

$$x^t[k+1] = Fx^t[k] + Bw^t[k] \tag{1}$$

$$y^t[k] = H^t[k]x^t[k] + v^t[k] \tag{2}$$

Where x is the target state vector, F is the dynamic model of the target, and H is the measurement model. k denotes a discrete time index, and t denotes a target index. The noise models w and v follow the covariance matrices Q_k^t and R_k^t as $w_k^t \sim N(0, Q_k^t)$ and $v_k^t \sim N(0, R_k^t)$ respectively.

The MTT problem can be divided into three problems: (1) M2TA; (2) track update; and (3) track management. The target measurements that the sensor acquires do not have an index of what the target is from, and there are also measurements by clutter that are not generated from the actual target. Therefore, the association between existing tracking and measurements is an important issue in tracking performance. Incorrect assignment of measurements to the track can lead to poor tracking performance or loss of track. In this paper, we use a relatively simple Global Nearest Neighbor (GNN) method. A description of GNN will be given in Sect. 2.3.

In addition to data association, track management is also important in MTT problems. Track management includes initializing tracks for newly discovered target measurements, deleting tracks without associated measures, and merging tracks when multiple tracks are generated for a target due to erroneous association.

2.2   KCF

KCF[1] is a well-known algorithm for deriving consensus on target state through exchange of information among multiple sensor nodes in a sensor network. One iteration of KCF is shown at Algorithm 1. Sensors exchange measurement data u_i, U_i and predicted track $\bar{x}_i$. Then, the fusion is performed using the measurements arrived from the neighboring nodes N_i. The target state is then updated through the consensus process. Where P_i and M_i are error covariances for prediction and estimated state information, respectively.

Algorithm 1. KCF

Initialize: x_i, P_i

1: Get measurement y_i
2: Compute information vector and covariance

$$\begin{cases} u_i = H_i R_i^{-1} y_i \\ U_i = H_i^T R_i^{-1} H_i \end{cases} \tag{3}$$

3: Broadcast message $m_i = (u_i, U_i, \bar{x}_i)$
4: Receive neighbor's messages $m_j, \forall j \in N_i$
5: Fuse information vector and covariance

$$\begin{cases} g_i = \sum_{j \in J_i} u_j \\ G_i = \sum_{j \in J_i} U_j \end{cases} \tag{4}$$

6: Compute the kalman-consensus state estimate

$$\bar{x}_i \leftarrow \bar{x}_i + M_i(g_i - G_i \bar{x}_i) + \gamma P_i \sum_{j \in N_i} (\bar{x}_j - \bar{x}_i) \tag{5}$$

$$M_i^t = ((P_i)^{-1} + S_i)^{-1} \tag{6}$$

$$\gamma = \epsilon/(\|P_i\| + 1) \tag{7}$$

7: Update local predicted state of targets

$$P_i^t \leftarrow F M_i F^T + B Q B^T \tag{8}$$

$$\bar{x}_i \leftarrow F \hat{x}_i \tag{9}$$

2.3   GNN

Among many of M2TA algorithms (JPDA [8,9], JIPIA [10], JITS [11], MHT [12], etc.), GNN is used in this paper because of it's simple applicability. GNN formulate M2TA as a problem of finding the permutation matrix M_{ij} of the following optimization problem.

$$\begin{aligned} \underset{M_{ij}}{\text{minimize}} \quad & \sum_{i=1}^{n} \sum_{j=1}^{m} C_{ij} M_{ij} \\ \text{subject to} \quad & \sum_{i=1}^{n} M_{ij} = 1, \\ & \sum_{j=1}^{m} M_{ij} = 1, \\ & M_{ij} \in 1, \end{aligned} \tag{10}$$

Here, C_{ij} has the following value. m and n is the number of measurements and tracks.

$$C_{ij} = \begin{cases} inf, & \text{if } d_{ij}^2 > G \\ d_{ij}^2, & \text{otherwise} \end{cases} \tag{11}$$

d_{ij}^2 is the innovation vector, defined as

$$d_{ij}^2 = \hat{r}_{ij}^T S_i^{-1} \hat{r}_{ij} \tag{12}$$

here,

$$\hat{r}_{ij} = y_j - H_i \bar{x}_i \tag{13}$$

$$S_i = R_i + H_i \bar{P}_i H_i^T \tag{14}$$

The GNN process is as follows.

(1) Gating:

$$\begin{cases} input: & \bar{x}^t, P^t, Y \quad \forall t \\ output: & Y^t, C \quad \forall t \end{cases} \tag{15}$$

The input and output of the gating function is (15). Where $Y = \{y^1, ..., y^m\}$ is the set of measurements and C is the cost matrix which is consist of (11). Y^t is the set of validated measurements for track t. Gating refers to excluding measurements that are less likely to be associated with tracks prior to M2TA by setting a probabilistic gate around the track. To do this, calculate d_{ij}^2 between the measurement and the track. Whether the measurement is in the validation gate is determined by the threshold value G.

$$d_{ij}^2 < G \tag{16}$$

After gating, each track will have one or more validated measurements. These measures are likely to be associated with track.

(2) Clustering:

$$\begin{cases} input: & Y^t \quad \forall t \\ output: & c^t, c^m \quad \forall t \, \forall m \end{cases} \tag{17}$$

c^t is the cluster to which the track t belongs, and c^m is the cluster to which the measurement value m belongs. A cluster is a unit group for computing an allocation algorithm. After clustering, tracks with the same measurements in the validation gate belong to a cluster. This process guarantee that one measurement is assigned to only one track at a time.

(3) M2TA (for all Clusters):

$$\begin{cases} input: & c^t, c^m, C \quad \forall t \, \forall m \\ output: & a^t \quad \forall t \end{cases} \tag{18}$$

Through the M2TA, the index of assigned measurement a^t for each track t is obtained. b is a set of measures that are not associated with any track. The linear assignment is performed for all unit clusters using C as the cost matrix. This gives a one-to-one match between measurements and tracks.

3   Track Management

The main purpose of track management is status (s^t) transitions of each track for correct situational awareness. In this process, the number of tracks is fixed to T, and it is necessary to distinguish whether these tracks are actually tracking the target. To do this, we use a track management method that includes track status update, track merge, and track initialization.

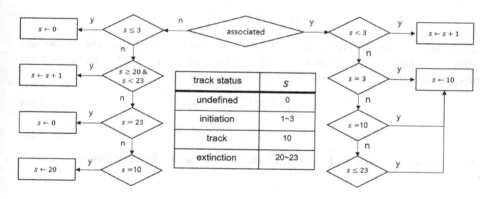

Fig. 2. Flow diagram of track status transition

(1) Track status update:

$$\begin{cases} input: & s^t, a^t, l^t, h^t, \bar{x}^t \quad \forall t \\ output: & s^t, l^t, h^t \quad \forall t \end{cases} \tag{19}$$

For track management, at this stage, performs a status update of the track depending on whether the track is associated with a measurement. s^t is status of track t. In this paper, we define four status for track management: undefined, initialization, track, and extinction. We use track status update logic of Fig. 2. l^t is the track's lifetime and h^t is the augmented track estimate for specific time interval e. l^t and h^t are used to merge tracks if multiple tracks generated from the same target cannot be deleted by themselves.

$$l_t = \begin{cases} l_t + 1 & \text{if } s^t \neq 0 \\ 0 & otherwise \end{cases} \tag{20}$$

$$h^t = \begin{bmatrix} \bar{x}^t[k - e + 1] \\ ... \\ \bar{x}^t[k] \end{bmatrix} \tag{21}$$

(2) Track merge:

$$\begin{cases} input: & \bar{x}^t, P^t, h^t, l^t, s^t \quad \forall t \\ output: & \bar{x}^t, P^t, s^t \quad \forall t \end{cases} \tag{22}$$

Track merging is necessary to delete unnecessary tracks when multiple tracks are created for one target. The metric used to delete a track can vary, and in this paper we use the similarity between the track estimate histories (h^t) as its metric. The similarity between tracks i and j is calculated as follows.

$$m_{ij} = ||h^i - h^j|| \tag{23}$$

If m_{ij} is less than specific threshold value, it is determined that the tracks are from same target. The following track to track fusion method is then used for merging between tracks i and j.

$$\bar{x} = P(P_i^{-1}\bar{x}_i + P_j^{-1}\bar{x}_j) \tag{24}$$

$$P = (P_i^{-1} + P_j^{-1})^{-1} \tag{25}$$

And then delete track with shorter lifetime from the tracks identified as the same track.

(3) Track initiation:

$$\begin{cases} input: & a^t, s^t, Y \quad \forall t \\ output: & \bar{x}^t, P^t \quad \forall t \end{cases} \tag{26}$$

Y is the measurement set. If there are unassociated measurements (which can be found with a^t), new track should be initialized with it.

4   GNN-KCF with Track Management

In this section, the GNN-KCF with track management is described (Algorithm 2). First, The number of tracks managed by the local sensor is assumed to be T. the measurements obtained from the local sensor are subjected to the GNN process. The M2TA result a_i^t and the unassociated measure set b_i are obtained.

Algorithm 2. GNN-KCF with Track Management for sensor i

Initialize: x_i^t, P_i^t $\forall t$

1: Do GNN based assignment :
   a: Gating between Y_i and $\bar{X}_i$
   b: Clustering
   c: M2TA $(a_i^t \leftarrow \{k|M_{tk} = 1\})$
2: Compute measurements for each track

$$\bar{y}_i^t = \begin{cases} y_i^{a_i^t}, & \text{if } \sum_{j=1}^m M_{ij} = 1 \\ H_i \bar{x}_i^t, & \text{otherwise} \end{cases} \tag{27}$$

$$d_i^t = \begin{cases} 1, & \text{if } \sum_{j=1}^m M_{ij} = 1 \\ 0, & \text{otherwise} \end{cases} \tag{28}$$

3: Compute information vector and covariance

$$\begin{cases} u_i^t = H_i R_i^{-1} \bar{y}_i^t, & \forall t \\ U_i^t = H_i^T R_i^{-1} H_i, & \forall t \end{cases} \tag{29}$$

4: Broadcast message $Msg_i = \{m_i^1, ..., m_i^T\}$, where $m_i^t = (u_i^t, U_i^t, \bar{x}_i^t, d_i^t)$
5: Receive neighbor's messages Msg_j, $\forall j \in N_i$
6: Do GNN based assignment $\forall j \in N_i$
   a: calculate $\bar{y}_j^{\,t}$

$$\bar{y}_j^t = \begin{cases} (H_i R_i^{-1})^{-1} u_j^t, & \text{if } d_j^t = 1 \\ NaN, & \text{otherwise} \end{cases} \tag{30}$$

   b: Gating between $\bar{Y}_j = \{\bar{y}_j^{\,1}, ..., \bar{y}_j^{\,T}\}$ and $\bar{X}_i$
   c: Clustering
   d: M2TA $(a_j^t \leftarrow \{k|M_{tk} = 1\})$
7: Track management $(a_i^t = \{a_j^t | j \in J_i\}, Y_i = \{\bar{Y}_j | j \in J_i\})$
8: Fuse information vector and covariance

$$\begin{cases} g_i^t = \sum_{j \in J_i} u_j^{a_j^t} \\ G_i^t = \sum_{j \in J_i} U_j^{a_j^t} \end{cases} \tag{31}$$

9: Compute the kalman-consensus state estimate

$$\hat{x}_i^t \leftarrow \bar{x}_i^t + M_i^t(g_i^t - G_i^t \bar{x}_i^t) + \gamma P_i^t \sum_{j \in N_i} (\bar{x}_j^t - \bar{x}_i^t) \tag{32}$$

$$M_i^t = ((P_i^t)^{-1} + S_i^t)^{-1} \tag{33}$$

$$\gamma = \epsilon/(\|P_i^t\| + 1) \tag{34}$$

10: Update local predicted state of targets

$$P_i^t \leftarrow FM_i^t F^T + BQB^T \tag{35}$$

$$\bar{x}_i^t \leftarrow F\hat{x}_i^t \tag{36}$$

The measurement associated with local track t is given in Eq. (27). If there is no associated measurement, use the predicted measurement. d_i^t indicates whether the transmitted message y_i^t is a measurement directly detected by the local sensor. d_i^t is used at the neighborhood nodes to fuse measurements from the actual measurements of the transmitter node. If this is not done, the erroneously generated track for a non-existent target can be maintained through communication between the sensors without being extinguished. Algorithm 2–6 is the GNN process between measurements from neighbor nodes and local tracks. Next, track management should be performed. The measurement set and association result of the measurement can be constructed by accumulation ($a_i^t = \{a_j^t | j \in J_i\}$, $Y_i = \{\bar{Y}_j | j \in J_i\}$).

5   Numerical Example

5.1   Simulation Setup

In order to verify the algorithms, sensor nodes with limited detection range are distributed sparsely. Nine distributed sensors track 13 targets and sensor detection range was set to 0.23. Constant velocity targets having an arbitrary initial position and velocity in a range of $0 < x < 1$, $0 < y < 1$, $0 < v_x < 0.05$, $0 < v_y < 0.05$ are randomly generated. The maximum number of tracks managed by the local sensor is set to be larger than the actual number of targets. The sensor model is as follows. We simulate a fully connected network and a network that is connected only between spatially close nodes (Fig. 3). Both cluttered and non-cluttered environment are considered. In a scenario with clutter, 17 clutters per time step are generated.

$$H_i = \begin{bmatrix} 1 & 0 & 0 & 0 \\ 0 & 0 & 1 & 0 \end{bmatrix} \tag{37}$$

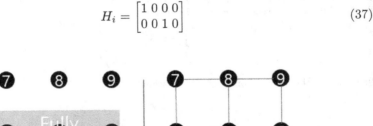

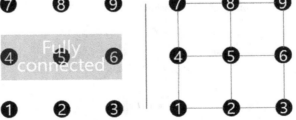

Fig. 3. Two network structures for simulation

5.2   Result

(1) Fully connected network

Figures 4 and 5 show the track of sensor 1 in a fully connected network. Since it is directly connected to all sensors in the network, it can be seen that sensor 1 has tracks for all targets in the environment.

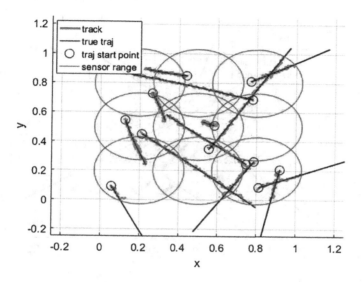

Fig. 4. Sensor 1 tracks (non-clutter) in fully connected networks

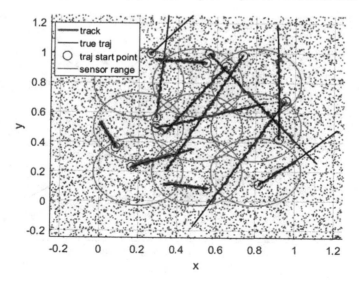

Fig. 5. Sensor 1 tracks (clutter) in fully connected networks

(2) Network only between close sensors

Figures 6 and 7 show the tracks of sensor 1 and sensor 9 in a network connected only between close nodes. As described in the algorithm, the data measured and transmitted from the sensor farther than the neighboring node ($N_1 \in \{2,4\}$) is not regarded as a valid measurement. Thus, sensor 1 has only tracks for targets that can be observed by sensors 1 and 2, 4.

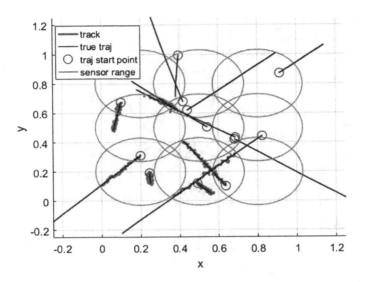

Fig. 6. Sensor 1 tracks (non-clutter) with network only with close sensors

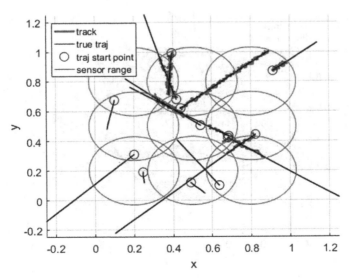

Fig. 7. Sensor 9 tracks (non-clutter) with network only with close sensors

6 Conclusion

Based on the KCF algorithm, we developed a multiple target tracking algorithm including M2TA and track management. To verify the algorithm, we constructed a scenario with limited sensor detection range and different FOVs. Numerical results show that track management works effectively in clutter and non-clutter environments.

Acknowledgements. This work was supported by Institute for Information & communications Technology Promotion (IITP) grant funded by the Korea government (MSIT) (No. 20150002130042002, Development of High Reliable Communications and Security SW for Various Unmanned Vehicles).

References

1. Olfati-Saber, R.: Distributed Kalman filtering for sensor networks. In: 2007 46th IEEE Conference on Decision and Control, pp. 5492–5498. IEEE (2007)
2. Wang, Z., Gu, D.: Cooperative target tracking control of multiple robots. IEEE Trans. Ind. Electron. **59**(8), 3232–3240 (2012)
3. Zhu, S., Chen, C., Li, W., Yang, B., Guan, X.: Distributed optimal consensus filter for target tracking in heterogeneous sensor networks. IEEE Trans. Cybern. **43**(6), 1963–1976 (2013)
4. Olfati-Saber, R., Jalalkamali, P.: Collaborative target tracking using distributed Kalman filtering on mobile sensor networks. In: 2011 American Control Conference (ACC), pp. 1100–1105. IEEE (2011)
5. Zhou, Z., Fang, H., Hong, Y.: Distributed estimation for moving target based on state-consensus strategy. IEEE Trans. Autom. Control **58**(8), 2096–2101 (2013)
6. Sandell, N.F., Olfati-Saber, R.: Distributed data association for multi-target tracking in sensor networks. In: 47th IEEE Conference on Decision and Control, CDC 2008, pp. 1085–1090. IEEE (2008)
7. Blackman, S.S.: Multiple-Target Tracking with Radar Applications, 463 p. Artech House Inc., Dedham (1986)
8. Bar-Shalom, Y., Willett, P.K., Tian, X.: Tracking and Data Fusion. YBS Publishing, Storrs (2011)
9. Bar-Shalom, Y., Li, X.R.: Multitarget-Multisensor Tracking: Principles and Techniques, vol. 19. YBS, London (1995)
10. Musicki, D., Evans, R.: Joint integrated probabilistic data association: JIPDA. IEEE Trans. Aerosp. Electron. Syst. **40**(3), 1093–1099 (2004)
11. Blom, H.A., Bloem, E.A.: Probabilistic data association avoiding track coalescence. IEEE Trans. Autom. Control **45**(2), 247–259 (2000)
12. Musicki, D., Evans, R.J.: Multiscan multitarget tracking in clutter with integrated track splitting filter. IEEE Trans. Aerosp. Electron. Syst. **45**(4), 1432–1447 (2009)

Deregulating the Electricity Market for the Peninsula Malaysian Grid Using a System Dynamics Approach

Darren Teck Liang Lee[1], Andrew Huey Ping Tan[2],
Eng Hwa Yap[3(✉)], and Kim-Yeow Tshai[1]

[1] Faculty of Engineering, University of Nottingham Malaysia,
Semenyih, Malaysia
{kedx6dlt, Kim-Yeow.Tshai}@nottingham.edu.my
[2] School of Engineering, Computing and Built Environment,
KDU Penang University College, Georgetown, Malaysia
tanandrew90@gmail.com
[3] Faculty of Transdisciplinary Innovation, University of Technology Sydney,
Sydney, Australia
EngHwa.Yap@uts.edu.au

Abstract. Competition among power producers is minimal in the current Peninsula Malaysia's electricity market structure as it only happens during the proposal stage to build new power plants. Therefore, electricity generation is over-dependent on current national gas reserves and the importation of cheap coal from Indonesia to keep tariffs low as there is minimal competition among power producers. However, this is not a suitable long term and environmental-friendly solution as it increases carbon dioxide and other greenhouse gases (GHG) emission. Hence, this research utilises system dynamics method to study the effects of a single buyer model (SBM) transitioning into a wholesale competition market, and to propose a framework to deregulate Peninsula Malaysia's electricity market. Results have shown that tariffs for end consumers are lower in the wholesale competitive market than the current SBM market even after incorporating expensive large-scale renewable energy.

1 Introduction

Privatisation of Malaysia's electric energy market has been ongoing since 1990 to allow independent power producers (IPPs) to generate electricity to meet the nation's increasing energy demand [1]. In an ideal deregulated market, generation and transmission of electricity must be divested to promote competition among IPPs. Currently in Malaysia, ownership in generation is already divested and the electricity market has already adopted a single buyer model (SBM). In this SBM arrangement, the national electricity company, Tenaga Nasional Berhad (TNB), is the single buyer responsible for purchasing electricity from power plants in Peninsular Malaysia and then selling directly to end consumers. The SBM is viewed as an intermediate solution before transitioning into a completely deregulated wholesale competition market. In the current SBM arrangement, competition among power plants are limited therefore

© Springer Nature Singapore Pte Ltd. 2019
J.-H. Kim et al. (Eds.): RiTA 2018, CCIS 1015, pp. 164–188, 2019.
https://doi.org/10.1007/978-981-13-7780-8_14

generation of electricity is dependent on current oil reserves and importation of cheap coals to keep tariffs affordable. Consequently, this will lead to an increase in carbon dioxide emissions which contributes to global warming. Therefore, there is a need to deregulate the electricity market further to reduce electricity tariffs for end consumers so that it will be more affordable to invest in large scale renewable energy to reduce greenhouse gases emissions.

2   Literature Review

2.1   Deregulated Electricity Market

Traditionally energy market is regulated and power plants are funded by public sector. Pricing are usually set by the government while the local state-owned utility have total monopoly over generation, transmission and distribution of electricity tasks. This structure is known as vertically integrated utility [2]. In the early 1990s, it was discovered that governments and even World Bank do not have sufficient funds to meet the increasing demand for electricity in developing countries [3]. Therefore IPPs were established in countries that require private finance to meet the growing electricity demand [4]. In an ideal deregulated energy market, ownership in generation and transmission of electrical energy by utilities must be fully divested and they are only responsible for [5]:

1. Distributing electricity, operations and maintenance of the grid to meter interconnection.
2. Billing consumers.
3. Acting as the last resort provider.

Benefits of a Deregulated Market and Privatisation of Electricity Supply Industry

Generally, a deregulated market is beneficial not only to the government but also to the public sector and environment. A list of benefits and reasons why countries chose to deregulate their electricity market is shown below:

1. IPPs from private capital markets have more funds than government to meet the country increasing demand of electrical power [6].
2. Reducing government expenditure and public debt problems [6].
3. Competitions between energy suppliers gave consumers a choice to choose an alternative energy supplier that provides cheaper rates [7].
4. Power producers compete with each other by increasing their productive efficiency in a deregulated market. Power producers will work at improving their technology and fuel efficiency [8].
5. Lower emissions of pollutants to the environment when power producers improved their technology and fuel efficiency [8].
6. Provides customer service for consumers [9].

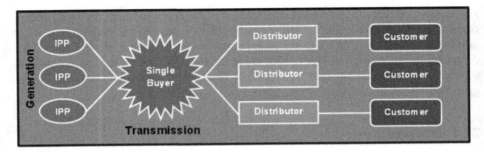

Fig. 1. Single buyer model [10]

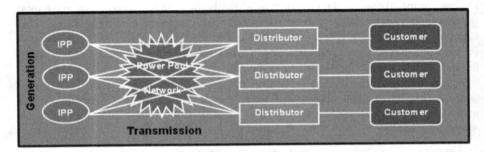

Fig. 2. Wholesale competition model [10]

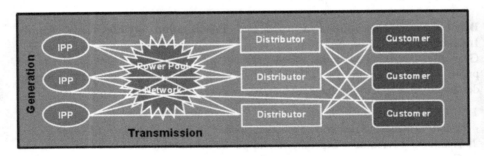

Fig. 3. Retail competition model [10]

Figures 1, 2, and 3 show general different types of deregulated ESI structure globally. The SBM shown in Fig. 1 is usually the first step of deregulation to introduce competition. The purchasing agency/single buyer is responsible for purchasing all the generated electricity from IPPs before selling it to distributor companies. In the wholesale competition model as shown in Fig. 2, distributor companies are given the choice to choose the IPPs they preferred to buy electricity from. This will lead to an increase in competition among IPPs. Meanwhile, the retail competition model shown in

Fig. 3 represents a fully deregulated ESI. Here, tariffs are no longer regulated so customers can choose their preferred energy retailer (distributor) that offers a better price.

Currently, TNB is the only distributor in Peninsular Malaysia. A retail competition market cannot be achieved unless Malaysia government permits new and private distributing companies into the market or unbundle TNB into several competing distribution companies.

2.2 Malaysia Current Market

Electricity Supply Industry (ESI) Structure

In mid 1990s, Malaysia has adopted SBM, as shown in Fig. 2, until now and IPPs were given licenses to generate electricity to TNB through power purchasing agreements (PPA) for 21 years [1]. This PPA includes a take-or-pay benefit which requires TNB to purchase electricity generated by IPPs even when there is no demand for it [11]. Presently, it has been replaced with second generation PPA where there is no longer take-or-pay benefit but includes bidding under Incentive-Based Regulation (IBR) to provide competitive tariffs for end consumers [11]. This means during the proposal stage to build new power plant, the second generation PPA will be awarded to IPP that has good company reputation and is able to bid the lowest levelised cost of electricity (LCOE) compared to the rest. Figure 4 below shows the current SBM structure where TNB will buy all electricity from IPPs and distribute to consumers.

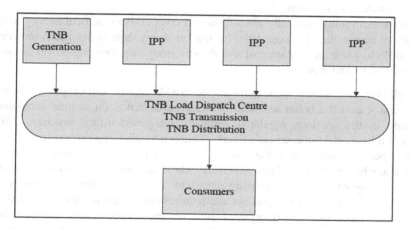

Fig. 4. Malaysia electricity supply industry [12]

There are several benefits to opt for single buyer model [13]:

1. Ease in balancing electricity supply and demand when there is only one party responsible for the transmission and distribution of electrical power purchased from power plant generators.

2. Lower cost because unlike multiple buyers and sellers model, there is little need for third-party access regime to transmission.
3. Price regulation are simplified due to a unified wholesale electricity tariff.
4. Power plant investors receives higher protection from market and regulatory risk.
5. This model may be appealing to policymakers who are reluctant to support a radically different competitive and deregulated wholesale energy market.

A single buyer model can be viewed as a transition to competitive wholesale energy market but evidence suggests that it is better to skip this stage and proceed to multiple buyers model immediately because the disadvantages listed below outweigh the advantages [13, 14]:

1. Government officials are not financially responsible for the consequences of their decision in increasing power generation capacity which might lead to over-investment.
2. There is a contingent liability for the government if national power company is unable to fulfil their obligation to IPPs.
3. Electricity prices rise instead of falling when demand falls short due to take-or-pay provision listed in the PPA where losses are passed on to taxpayers.
4. Hinder cross-border electricity trade development as the national power company does not have a strong motive in making profits.
5. The incentives for distributors in collecting payments from customers are weaker because the national power company are reluctant in taking politically unpopular action against a delinquent distributor.
6. Government can intervene easily.
7. Delay transitioning into fully liberalized and competitive electrical energy market.
8. Without competition between IPPs for market share, there is not much motivation for IPPS to lower generation cost and the efficiency of generating power are solely motivated to make profit.

Although the mentioned disadvantages of SBM clearly outweigh its benefits, nonetheless it is still a better solution for Malaysia at that point in time compared to remaining in the previous regulated vertically integrated utility structure. This is because it relieves the government's financial burden by allowing private sector to fund and build power plants to meet the growing electricity demand. On the other hand, the market is unable to be deregulate further into wholesale competition model due to the country lacking experts in liberalizing electricity market. Furthermore, deregulating electricity market is a complex process and requires substantial research and studies to minimize complication. Therefore, until there are more experts and studies of liberalizing electricity market carried out, SBM is currently the most suitable intermediate solution for Malaysia.

Malaysia Fuels for Electricity Generation
Figures 5 and 6 shows that oil is no longer the main source of fuel to generate electricity and it has been replaced by coal while the growth of renewable energy is slow even after 22 years. In addition, natural gas is still shown to be the main fuel for electricity generation.

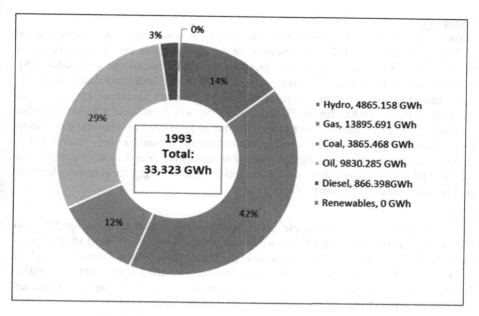

Fig. 5. Malaysia electricity generation mix in 1993 [15]

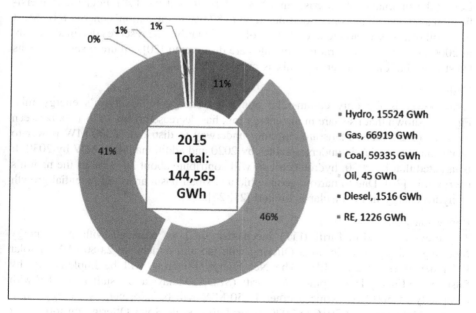

Fig. 6. Malaysia electricity generation mix in 2015 [16]

Natural Gas

Malaysia has the second largest natural gas reserve in South East Asia and ranked the 12th largest in the world [17]. The reserves are mainly located in the shelves offshore of Sabah and Sarawak while some are found in the east coast of the peninsular [18]. Although an investigation was conducted in 2010 which indicated that natural gas could only last for additional 36 years as Malaysia's main source of energy, it was reported by PETRONAS that the oil and gas reserve will decrease at a higher rate after a decade [19]. Hence the country cannot continue to just depend on natural gas reserve for its future energy needs in meeting the increasing demand unless natural gas is imported from overseas. Even though renewable energy has been introduced as the fifth fuel to Malaysia's energy mix, the consumption of natural gas to generate electrical power still increases as shown in Fig. 6.

Coal

There are abundant coal reserves in Sabah and Sarawak which accounts for 29% and 69% of the energy mix respectively while the other 2% are only found in the peninsular. However these reserves are located in the inland areas which have severe lack of infrastructure for extraction [18], therefore coal are imported from China, Australia, Indonesia and South Africa [17]. Since coal is one of the cheapest fuel, it is expected that natural gas will slowly be replaced by coal in the future [18].

Oil

Before the international oil crisis in 1973 and 1979, and the 1981 Four-fuel Diversification strategy, oil has once contributed up to 87.9% in the nation's energy mix. Most of the oil reserves are found in east coast of peninsular but the reserves declined sharply in 2006 even though several new oil sites are discovered [20]. Figure 6 shows that its latest contribution to the energy mix is only at 1%.

Hydro

Hydropower is the only commercial non-fossil energy in Malaysia's energy mix. Figure 6 shown that its share in the energy mix has decreased from 14 to 11% between the gaps of 20 years. Plans are currently underway to distribute 5000 MW power to peninsular Malaysia via undersea cables by 2020 and additional 5000 MW by 2030. It is targeted that by 2030, hydroelectricity will contribute about 30–35% of the nation's energy mix [20]. Due to narrow geographic and small basin area, the potential growth of hydro energy in Peninsular is limited [21, 22].

Renewables

The previous Feed-in-Tariff (FiT) mechanism is now replaced with Net Energy Metering and Large Scale Solar Offering with the aim to achieve 2,080 MW of solar PV renewables by 2020 [23]. The Net Energy Metering will be implemented by Sustainable Energy Development Authority (SEDA) Malaysia to install a solar PV with a capacity of 500 MW while another 1,250 MW will be allocated by Energy Commission Malaysia from 2016 to 2020 under Large Scale Solar Offering on top of our existing solar PV capacity [24]. Solar PV is the preferred choice among other renewables such as small hydropower, biogas and biofuel because it has the highest potential source among them to meet electricity demand [25]. 15 According to a study

[26], the average annual energy output of rooftop solar PV application in Malaysia is among the top half when compared to 41 cities from 26 countries.

2.3 Research Rationale

There are only a handful of available researches related to liberalizing Malaysia's electricity market. The most relevant studies available to date are from Hassan et al. [27] and Arifin [28], both studies have determined which wholesale competition model is suitable to deregulate Malaysia's SBM further. Figures 7 and 8 below show the different types of wholesale competition model where IPPs have to compete with other through bidding. There is another model known as hybrid model which is a combination of pool and bilateral contract model.

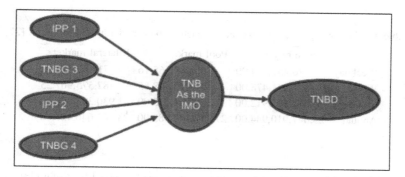

Fig. 7. Single auction power pool model [27]

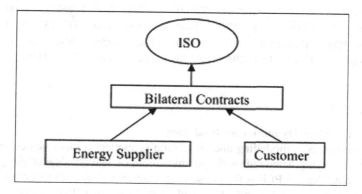

Fig. 8. Bilateral contract model [27]

Both Hassan et al. and Arifin determined and analysed the overall revenue of IPPs by using MATLAB simulation in their studies. Table 1 shows the results of Hassan et al.'s research. The pool market cost more than single buyer due to uniform price scheme used in the study, however it was suggested that this amount could be reduced with the implementation of a suitable policy to increase TNB's savings.

The results of Arifin's research are shown in Table 2. Hybrid structure (Base_PAB and Base_UP) and pool (Pool_PAB and Pool_UP) cost less than SBM so it will increase TNB's savings, which can be passed on to end consumers. Both studies concluded that SBM is not sustainable for long term and there is a need to introduce competitive market structure to reduce tariffs for end-consumers.

In both studies, there was no framework developed to help transition the current SBM arrangement into a wholesale competitive model. Moreover, the results of both studies focused on the revenue of power producers instead of tariffs for end consumers. Hence this present research aims to fill this gap while also study the market correlation with renewable energy and its impact on tariffs by simulating different electricity market scenarios for 40 years.

Table 1. The total generation revenue for each market model by Hassan et al. [27]

	Single buyer	Pool market	Bilateral market
Weekday	76,457,064.00	86,080,665.16	71,167,526.92
Week	521,831,478.00	577,440,428.85	482,575,967.25
Month	2,087,325,912.00	2,309,761,715.00	1,930,303,869.00
Annual	25,047,910,944.00	27,717,140,585.00	23,163,646,428.00

Table 2. The total generation revenue for each market model by Arifin [28]

	Weekday	Week	Month	Annual
SBM	56084902.66	383680394.60	1534721578.48	18416658940.00
Pool_PAB	40764836.00	276439928.00	1105759712.00	13269116540.00
Pool_UP	48989240.00	327899020.00	1311569080.00	15739152960.00
Base_PAB	41868559.41	285212567.60	1140850270.56	13690203240.00
Base_UP	43403681.22	294609863.50	1178439454.04	14141273450.00

3   Methodology

Overview of System Dynamics Methodology

System dynamics (SD) modelling and simulation is applied in this research due to the complex time-varying behaviour and systemic nature of Malaysia's electricity market. In this research, Vensim PLE is the chosen software to perform SD modelling. In SD, Causal loop diagram (CLD) and Stock and Flow Diagram (SFD) are the two main models to describe complex system.

The purpose of CLD is to help identify causal relationship among important variables identified through literature review. In addition, it acts as a mental model to portray the boundaries of a system. Furthermore, readers who are unfamiliar with system dynamics methodology will be able to understand the behaviour of a particular complex system better since the interaction among variables are shown with using simple symbols such as arrows and polarity. A positive polarity indicates that when an independent variable increase (or decrease), the dependent variable increases (or decreases) as a result. As for negative polarity, while the independent variable decreases (or increases), the dependent variable increases (or decreases). In CLD there are two different feedback loops and they occurred when variables are connected in a circular manner. Below are the explanation for these two feedback loops [29]:

Reinforcing loops:

- In a reinforcing loop, results from an action will influence more of the same action which causes growth or decline of the results in an increasing rate.
- Reinforcing loop happens when the impact of a change increases by feedback.
- Virtuous cycles are produced by positive reinforcing loops while vicious cycles are produced by negative reinforcing loops.
- Balancing loops generate resistance to limit growth, maintain stability, and achieve equilibrium.
- Impact of a change is reduced by balancing loops.

A completed CLD model can be reconstructed into SFD model. This process is much easier as compared to developing SFD model from scratch because the later requires identifying which variable is level and rate while identifying the causal relationship among them simultaneously. Unlike CLD, SFD model contain more detailed information such as levels, rates and mathematical formulae. Simulation is carried out in SFD to determine numerical changes and behaviour analysis with the influence of time. In this research, numerous simulation run will be performed with different parameter to replicate different scenarios of Peninsular Malaysia's electricity market.

Figure 9 is the CLD diagram showing the dynamics of Peninsular Malaysia's electricity market. The blue, red and green colour area in the diagram represents scenarios 1, 2 and 3 respectively. For scenario 1, Loop B1 and B2 shows that consumers will use more electricity when tariffs are low. A growing GDP consumes more electricity hence it will increase electricity demand. The consumption rate of fossil fuel will increase to meet the demand hence leading to more coal imports and increase depletion rate of natural gas reserve. As a result, the overall fossil fuel price to generate electricity increases hence increasing the tariffs for consumers. Loop R1 shows that new second generation PPA power plants are built to meet the increasing demand for electricity. At the same time when first generation PPA of individual power plants expired, they will be renewed with second generation PPA. Since the first generation PPA contains take or pay quota [11] which requires TNB to purchase electricity from power plants even when demands fall, this will increase tariffs for consumers. By referring to literature review section, first generation PPA was no longer issued out hence it is replaced with second generation PPA which contains only energy and capacity quota [11] to reduce tariffs.

In the scenario 2, despite having second generation PPA tariffs continue to increase as the market is still in the single buyer model. Thus, third generation PPA is proposed to replace the existing PPA and promote competitive bidding hourly among power plants to sell TNB electricity. Due to this bidding, the market is transitioning from SBM into wholesale competition market. Tariffs will be reduced gradually due to retirement of older power plants under first and second generation PPA and also due to bidding quota under third generation PPA. A complete wholesale competition market will be achieved when all the first and second generation PPA power plants retired.

As scenario 3, it is expected that CO2 emissions increases when the number of non-RE power plants increases. Therefore, there is a need to promote RE growth to address these issues. To promote RE growth, current Net Energy Metering and Large-Scale Solar Offering need adjustments to make it more rewarding than the third generation PPA. As RE power plants require higher investment, tariffs for end consumers will increase. This behaviour is shown in loop R3 and B4.

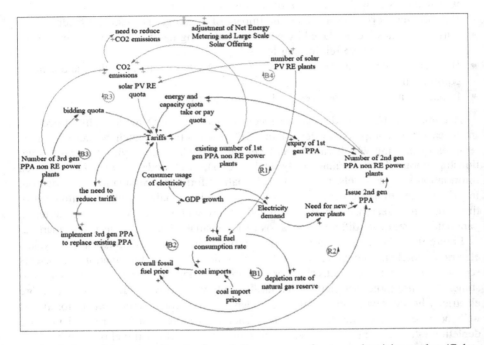

Fig. 9. CLD model representing the deregulation process of current electricity market (Color figure online)

Figures 10, 11 and 12 show SFD models that are reconstructed from Fig. 9 CLD to carry out simulations from year 2013 to 2053. Table 3 below shows three different scenarios for Peninsular Malaysia electricity market. Scenario 1 shown in Fig. 10 model represents the current SBM structure of Peninsular Malaysia. In scenario 2 shown in Fig. 11, the model is built based on Fig. 9 but introduced a third generation PPA to help transition the market from SBM structure into wholesale competition structure. The variable "issue of second generation PPA" in scenario 2 will be running at first but replaced by third generation PPA when the need to reduce tariffs for consumers became more urgent. The need is determined by the threshold value shown in Table 4. Unlike first and second generation PPA, power plants licensed with third generation PPA are required to bid hourly to sell electricity to TNB among each other.

In scenario 3 shown in Fig. 12, additional variables related to RE growth are included as an expansion of scenario 2. With the increasing amount of carbon dioxide emission, RE capacity will increase as a result of adjustment of Net Energy metering and large-scale solar offering policies to address this issue. Both policies are adjusted so that solar PV RE plants do not need to compete with third generation non-RE plants to sell TNB electricity to encourage RE growth. At the same time, the growth of RE capacity is controlled to keep tariffs affordable for end consumers since the cost of investing in solar PV is high. This adjustment also means that the building of solar PV plants is given more incentives to be made more rewarding than building natural gas and coal power plants.

To make modelling simpler and due to the fact Malaysia is increasingly dependent on coal imports for electricity, hydro is not added into all SFD models even though it contributed 11% to Malaysia's 2015 electricity generation mix shown in Fig. 6. Moreover, the potential growth of hydropower is limited in Peninsular Malaysia due to narrow geographic and moderately small river basins [21, 22]. As for renewable energy, its initial capacity is set to 0 in scenario 3 as it only contributed 1% to electricity generation mix shown in Fig. 6.

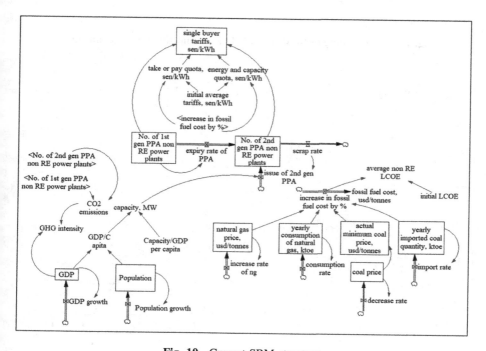

Fig. 10. Current SBM structure

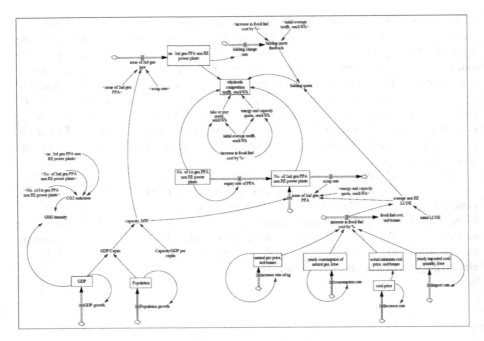

Fig. 11. Transitioning of SBM into wholesale competition market structure

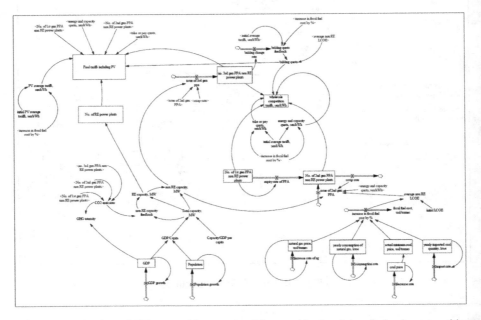

Fig. 12. Introduction of PV renewable energy while transitioning into wholesale competition market structure

Table 3. Description summary of different scenarios for simulation

Scenario	Description
1 Current Peninsular Malaysia SBM electricity market (Fig. 10)	This is the current scenario for Malaysia SBM electricity market. Power plants under first generation PPA are paid with take-or-pay quota whereas power plants under second generation PPA are paid with energy and capacity quota. The take-or-pay quota are more profitable but will increase the tariff for end consumers
2 Introduction of third generation PPA (Fig. 11)	The model for this scenario introduced a new third generation PPA variable to replace the existing first and second generation PPA so that the market can transition from SBM into wholesale competition market. Power plants under third generation PPA have to bid actively among each other every hour to sell electricity to TNB
3 Introduction of large-scale PV renewable growth (Fig. 12)	This is an extended model of scenario 2 to introduce both third generation PPA and PV renewable energy

4 Results and Findings

Scenario 1: Current Peninsular Malaysia SBM Electricity Market

The results of Figs. 13 and 14 represents the current SBM electricity market of Peninsular Malaysia. Figure 13 contains the data and trend for take or pay quota, energy and capacity quota. As mentioned in Tables 2 and 3, the take or pay quota is influenced by power plants under first generation PPA while the energy and capacity quota is influenced by power plants under second generation PPA. Both quotas influenced the final tariff of SBM. The current second generation PPA currently replaces the first generation PPA so that newer power plants will no longer have take-or-pay conditions are only paid by energy and capacity payments. Therefore, the number of first generation PPA power plants will drop in the beginning whereas the number of second generation PPA power plants will only increase in the future. This was intended to increase competition to reduce tariffs for end consumer through first bidding exercise in IBR. Despite the implementation of second generation PPA, the final tariffs were only 13.28% lower than the first generation PPA take or pay quota at the end of simulation. Figure 13 shows that the final tariff is at first same price as take or pay quota. As the number of second generation PPA power plants increases, the final tariff begins to reduce eventually until it is the same value as energy and capacity quota.

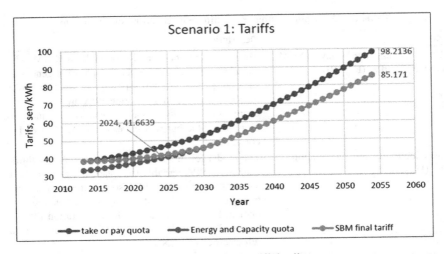

Fig. 13. Scenario 1 tariff details

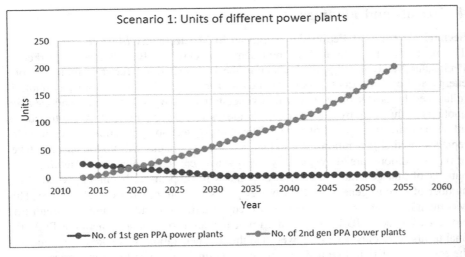

Fig. 14. Different types and number of power plants in scenario 1

Scenario 2: Introduction of Third Generation PPA

In scenario 2, the third generation PPA is introduced in 2032 to replace the existing first and second generation PPA so that the market can transition from SBM into a wholesale competition market. Power plants under third generation PPA are required to bid hourly to sell electricity to TNB. Therefore, the wholesale competition tariff is influenced by take-or-pay (first generation PPA), energy and capacity (second generation PPA) and bidding (third generation PPA) quota. Figure 15 shows that after introducing third generation PPA, the bidding quota lowered the wholesale competition tariff for end consumers.

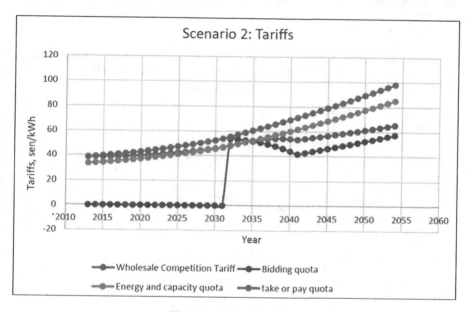

Fig. 15. Scenario 2 tariffs

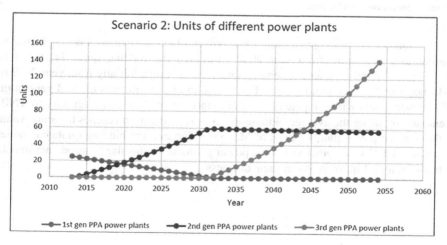

Fig. 16. Different types and number of power plants in scenario 2

Figure 16 shows the different types and number of power plants. In year 2032, third generation PPA is implemented to replace second generation PPA. The number of second generation PPA power plants became stagnant and gradually decreases from year 2032 whereas third generation PPA power plants grow exponentially to meet the rising electricity demands of Peninsular Malaysia. Originally the second generation PPA is used to replace first generation PPA in year 2014 therefore the number of first generation PPA power plants declined from beginning of the simulation and reaches 0 at year 2032.

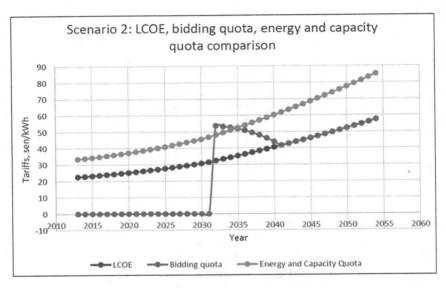

Fig. 17. Comparison of LCOE, bidding quota (third generation PPA) and energy and capacity (second generation PPA) quota

The number of third generation PPA power plants shown in Fig. 16 increases continuously beginning from year 2032. Although the bidding quota is shown in Fig. 17 to be slightly higher than energy and capacity quota initially from year 2032 to 2034 but it will reduce rapidly later. This is due to higher numbers of third generation PPA power plants will compete and bid among each other to sell electricity to TNB, eventually reducing the average bidding quota as shown until it reaches the same value as the average power plants LCOE of Malaysia. The average bidding quota can never go below LCOE value otherwise most power plants in Peninsular Malaysia are unable to breakeven and make profit in their business.

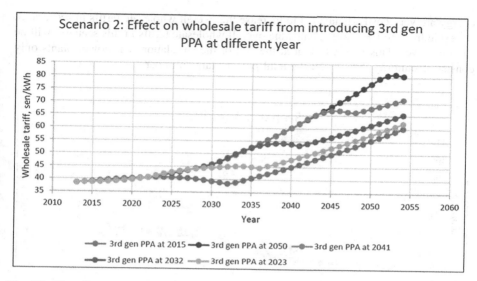

Fig. 18. The effect on wholesale competition tariffs when third generation PPA are introduced at different years

Table 4. Implementation of third generation PPA at different threshold values

Year to implement third generation PPA	Threshold value "Differences in second generation (energy & capacity) and average non RE LCOE quota, sen/kWh"	Wholesale tariffs at the end of simulation, sen/kWh
2015	10.0	60.16
2023	12.5	62.11
2032	15.0	65.40
2041	20.0	71.30
2050	25.0	80.63

The results in scenario 2 shown previously from Figs. 15, 16 and 17 are from implementing third generation PPA in the year 2032; when the differences in second generation and average non-RE LCOE quota became 15 sen/kWh. This value is assumed to be the threshold where the differences are too large for government and Peninsular residents to accept the higher price of final SBM tariffs, hence the urgent need to introduce third generation PPA to shift the SBM market into wholesale competition to reduce tariffs for end consumers. Table 4 shows that if this threshold value is higher, then the third generation PPA will be implemented later on. Figure 18 and Table 4 shows that the final wholesale competition tariffs for end consumers will be lower if third generation PPA is introduced earlier.

Scenario 3: Introduction of Large Scale RE Growth
In scenario 3, PV plants are installed to promote renewable energy growth and reduce carbon dioxide emissions. Figure 19 show that the average tariffs of PV renewable

energy are very costly, hence led to increase the overall wholesale tariffs in scenario 3. By excluding the RE power plants, the overall wholesale tariffs in this scenario will be slightly lower. This is due to that at the end of the simulation, RE power plants only consist of minority 20% of all the total power plants number.

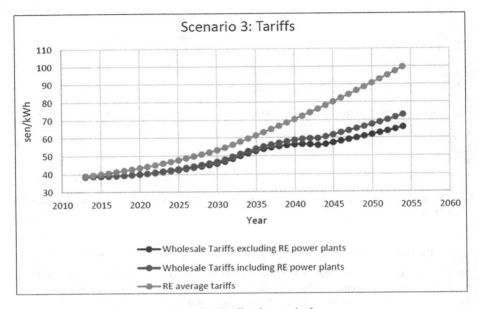

Fig. 19. Tariffs of scenario 3

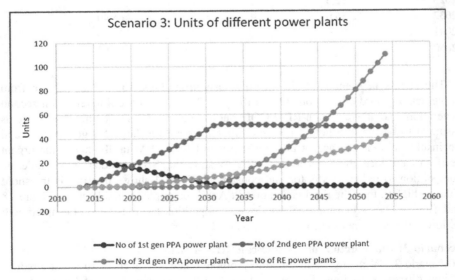

Fig. 20. Different types and number of power plants in scenario 3

Figure 20 shows the number of different types of power plants in scenario 3. As the demand for electricity increases due to growing GDP and population in peninsular Malaysia, more natural gas and coal power plants are built to meet the electricity demand. This will result in an exponential increase of CO_2 emissions as shown in Fig. 21 below. Hence the number of RE power plants increases as shown in Fig. 20 to address this issue.

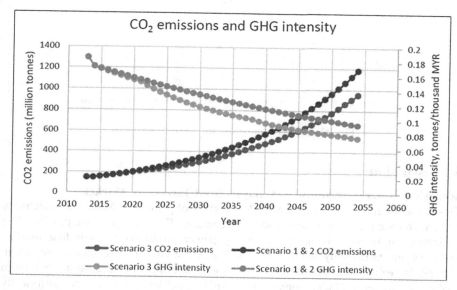

Fig. 21. Comparison of CO_2 emissions and GHG intensity among all scenarios

Comparison of All Scenarios

Figure 21 compares the CO_2 emission and GHG intensity in all the simulated scenarios. In the first two scenarios, the number of power plants and electric capacity is the same therefore they have the same values of CO_2 emissions and GHG intensity. By introducing large scale PV renewable energy into scenario 3, both emissions of CO_2 and GHG intensity reduced by 20%.

Figure 22 shows that the tariffs in scenario 2 is the lowest among all. The tariffs in scenario 2 and scenario 3 is lower than scenario 1 by 23% and 14.4% respectively. In all three scenarios, third generation PPA is implemented at year 2032, which is when the differences in second generation (energy & capacity) and average non-RE LCOE quota reached 15 sen/kWh. Therefore, the differences in tariffs among all 3 scenarios only begin to diverge shortly after year 2032.

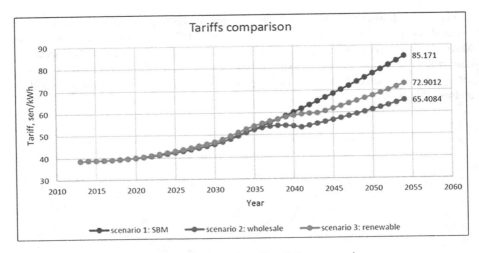

Fig. 22. Tariffs comparison for all three scenarios

5 Discussion

Scenario 1: Current Peninsular Malaysia SBM Electricity Market

This scenario represents the current SBM structure of peninsular Malaysia's electricity market. Despite the introduction of second generation PPA to replace the previous first generation to increase competition among power producers through first bidding exercises, the tariffs only reduced by 13.28% at the end of simulation as shown in Fig. 19. In reality, the final tariffs will be higher than Fig. 11 because the minimum coal price was set to USD30/tonnes in the SFD model when it was never historically lower than USD50/tonnes [30]. This is to show that there is a need to restructure and further deregulate the electricity market because of the rising tariffs despite taking advantage of importing cheap coals. Currently, the government uses PPA saving funds to absorb the rising surcharge in tariffs through ICPT rebates in order not to burden Peninsular Malaysia consumers [31] but there is no guarantee this funds can act as a long term solution. Surcharge in tariffs are not only caused by global inflation but also due to currency exchange; weaker MYR against USD led to higher cost of importing coal. In short, the current SBM structure is not a sustainable long-term solution for the electricity market of Peninsular Malaysia.

Scenario 2: Introduction of Third Generation PPA

In this scenario, third generation PPA is introduced to replace all existing PPA to deregulate the current SBM electricity market further into wholesale competition model. A stakeholder engagement was conducted with a subject expert researcher in Malaysia's electricity industry to obtain feedback for the findings of this research and insights for the current market. The feedback given was good and a suggestion was made by the expert to determine how the final wholesale competition tariff will be affected by introducing third generation PPA at different time period. Thus, Fig. 18 and Table 4 are plotted and tabulated in response to this suggestion. It was shown that by implementing third generation PPA earlier, the wholesale competition tariffs will be lower to the benefits of peninsular consumers. Meanwhile, Fig. 17 show that the bidding quota of third generation PPA payment surged higher than second generation PPA payment initially but this effect is only temporary. This is due to the market adjusting itself as it only had a few power plants under the new third generation PPA but after that the wholesale competition tariffs will quickly reduce until it became lower than energy and capacity quota as shown in Fig. 17. Higher number of power plants under third generation PPA will reduce the average wholesale competition tariff until it eventually became the same as average LCOE. If the tariffs is not controlled and allowed to be lower than the average LCOE then most of the third generation PPA power plants will be making a loss in their business.

Scenario 3: Introduction of Large Scale RE Growth

Renewable energy specifically solar PV is added into this scenario. Currently Malaysia has set a target to achieve 40% reduction in greenhouse gases by 2020 [32]. Below is the current formula to calculate the reduction in Malaysia greenhouse gases [33],

$$GHG\ intensity = \frac{CO_2\ emitted\ (tonnes)}{DP\ of\ Malaysia\ (1000\ MYR)}$$

However, this formula is not a good representation to determine the progress of CO2/GHG reduction. Figure 21 shows that despite the reduction in GHG intensity, the amount of CO2 emitted annually is growing exponentially.

As mentioned in the literature review section, both policies of Net Energy Metering and Large Scale Solar Offering planned to installed additional 1,750 MW of solar PV capacity from year 2016–2020 [24]. Since the simulation begin from year 2013, the average solar PV capacity to be installed is 250 MW a year. In fact, this is actually very little because at the end of simulation, the total capacity needed to meet the exponential growth of GDP and Peninsular population electricity demand is 197,306 MW. Therefore, both of these solar policies are adjusted drastically to reduce CO2 emissions while meeting the ever-increasing demand for electricity in scenario 3 model shown in Table 5 below.

Table 5. Adjustment of net energy metering and large-scale solar offering to increase solar PV capacity to reduce CO2 emissions

CO2 emissions (million tonnes)	Allocated annual capacity for solar PV, MW
Less than 200	250
Between 200 and 400	700
Between 400 and 800	1500
Above 800	2500

Figure 21 shows that despite making the drastic adjustment to both solar policies (known as RE capacities in Fig. 11 model), the emissions of CO2 continue to increase exponentially. Nevertheless, the emissions are 20% lowered compared to scenario 1 &

2. As long the number of RE power plants increases exponentially as shown in Fig. 20 then the gap differences of CO_2 emissions between non-RE (scene 1 & 2) scenario and scene 3 will continue to widen significantly as shown in Fig. 21 even after 40 years.

By analysing Fig. 22, scenario 1 represents the current SBM Peninsular electricity market and its tariff is 23% and 14.4% higher than scenarios 2 and 3 respectively. Scenario 2 provides the cheapest tariff option for Peninsular consumers because power producers under third generation PPA must bid competitively among each other sell electricity to TNB. By introducing renewable energy into scenario 3, the tariff became 11.45% higher than scenario 2 since the installation of large-scale solar PV requires high investment cost. Moving into a more competitive deregulated market shown in scenario 2 will not just reduce tariff for consumers but also give them and government a choice whether or not to take advantage of the lower tariffs and opt for scenario 3 although it will raise the tariffs a little higher. In other words, staying at the current scenario 1 SBM market will be more difficult to promote solar PV renewable energy growth as the tariffs were already high to begin with.

6   Conclusion

In conclusion, this research has developed a framework for deregulating Peninsular Malaysia's SBM electricity market into a wholesale competition model. From scenario 1 – current SBM, first generation PPA are replaced with second generation PPA so power producers must compete with each other via first bidding exercise during the proposal stage to build new power plants. Through this replacement, power plants under second generation PPA no longer have the benefit of take-or-pay condition so tariffs for consumers will be lower. However, the results of this scenario show that the final tariffs will only be lower than first generation PPA payment by 13.28% at the end of simulation period. Therefore, third generation PPA is introduced into scenario 2 replacing all the existing PPA so that power plants will bid competitively with each other to sell their generated electricity to TNB. As a result, the overall tariffs will be reduced by 23% compared to the final tariff of scenario 1. Once all the power plants under first and second generation PPA have retired, the electricity market will be in full wholesale competitive mode. On top of that, the results of scenario 2 show that tariffs will be lowered if third generation PPA is introduced earlier. In the scenario 3, large scale solar PV renewable is introduced as an expansion of scenario 2. The simulated tariff of this scenario is higher than scenario 2 by 11.45% and lower than scenario 1 by 14.4%. Since the tariffs in scenario 2 is significantly lower than scenario 1, this will put Malaysia in a better position to opt for scenario 3 and invest in renewable energy although it will raise the tariffs for the end users. In addition, investing in renewable energy can be made more affordable by introducing third generation PPA earlier. If the current market structure shown in scenario 1 stays the same without deregulating any further, then public acceptance and the government's willingness to invest in renewable energy might be lower because it will increase scenario 1's tariffs which is already high compared to scenario 2. In short, the SBM structure shown in scenario 1 is not a sustainable long-term solution to promote the growth of renewable energy and to reduce tariffs for end users; therefore, there is a compelling need to deregulate the electricity market further.

References

1. Malaysia Energy Commission: Peninsular Malaysia Electricity Supply Industry Outlook (2013)
2. Borenstein, S., Bushnell, J.: The U.S. electricity industry after 20 years of restructuring, 1–31 (2015). Energy Institute HAAS. https://doi.org/10.1146/annureveconomics-080614-115630.1
3. Strickland, C.: Energy efficiency in World Bank power sector policy and lending New opportunities. Energy Policy **26**, 873–883 (1998). https://doi.org/10.1016/S0301-4215(98)00007-X
4. McGovern, T., Hicks, C.: Deregulation and restructuring of the global electricity supply industry and its impact upon power plant suppliers. Int. J. Prod. Econ. **89**, 321–337 (2004). https://doi.org/10.1016/j.ijpe.2004.03.006
5. Woodcock, R.: Regulated and deregulated energy markets, explained. In: Energy Smart (2014). http://www.energysmart.enernoc.com/regulated-and-deregulated-energy-markets-explained/
6. Jacob, F.X.: Privatization and regulation in Malaysia's power sector. Energy Sustain. Dev. **3**, 80–88 (1997)
7. Oasis Energy Deregulation Explained - Oasis Energy. https://www.oasisenergy.com/index.php/energy-resources/deregulation-explained. Accessed 12 July 2017
8. Lefevre, T., Todoc, J.L.: Energy Deregulation in Asia : Status, Trends, and Implications on the Environment by Overview of Energy-Environment Situation, pp. 1–25 (1999)
9. Foo, K.Y.: A vision on the opportunities, policies and coping strategies for the energy security and green energy development in Malaysia. Renew. Sustain. Energy Rev. **51**, 1477–1498 (2015). https://doi.org/10.1016/j.rser.2015.07.041
10. Teljeur, E., van der Hoven, Z., Kagee, S.: Why we need alternative ESI structures. In: The Electricity Retail Competition Unicorn, 2nd Annual Competition and Economic Regulation (ACER) Conference, Southern Africa (2016)
11. Omar, A., Omar, O., Mohd Nazri, A.A., et al.: Energy Malaysia. Energy Comm. **1**, 48 (2014)
12. Kasim, N.N.B.: Study on Single Buyer Model and Pool Trading Model in Deregulated Electricity Market, p. 54 (2014)
13. Laszlo, L.: The single-buyer model. Privatesector **225**, 4 (2000)
14. Bacon, R.: Competitive contracting for privately generated power: what to do in the absence of competition in the market **47**, 1–4 (1995)
15. Energy Commission: Malaysia Energy Statistics Handbook 2015, p. 84 (2015)
16. Energy Commission: Malaysia Energy Statistics Handbook 2016 (2016)
17. Rahman Mohamed, A., Lee, K.T.: Energy for sustainable development in Malaysia: energy policy and alternative energy. Energy Policy **34**, 2388–2397 (2006). https://doi.org/10.1016/j.enpol.2005.04.003
18. Ali, R., Daut, I., Taib, S.: A review on existing and future energy sources for electrical power generation in Malaysia. Renew. Sustain. Energy Rev. **16**, 4047–4055 (2012). https://doi.org/10.1016/j.rser.2012.03.003
19. Economic Transformation Programme: Powering the Malaysian Economy with Oil, Gas and Energy, chapter 6. Economics Transformation Program. A Roadmap Malaysia, p. 27 (2010)

20. Oh, T.H., Pang, S.Y., Chua, S.C.: Energy policy and alternative energy in Malaysia: issues and challenges for sustainable growth. Renew. Sustain. Energy Rev. **14**, 1241–1252 (2010). https://doi.org/10.1016/j.rser.2009.12.003

21. Hydroelectric Power: Hydropower Scenario in Malaysia (2014). http://hydroelectricp.blogspot.my/2014/12/hydropower-scenario-in-malaysia.html. Accessed 24 Aug 2017

22. Ho, Y.J.: Out of gas weighing our energy options. In: Penang Mon (2014)

23. Lee, J.: A growing solar industry. Star Online (2017)

24. Kenning, T.: PV Talk: the policies needed to drive Malaysian solar. In: PVTECH (2016)

25. Chua, S.C., Oh, T.H.: Solar energy outlook in Malaysia. Renew. Sustain. Energy Rev. **16**, 564–574 (2012). https://doi.org/10.1016/j.rser.2011.08.022

26. MBIPV: Compared Assessment of Selected Environmental Indicators of Photovoltaic Electricity in Selected OECD Cities and Malaysian Cities, p. 20 (2006)

27. Hassan, M.Y., Hussin, F., Othman, M.F.: A Study of Electricity Market Models in the Restructured Electricity Supply Industry, p. 168 (2009)

28. Arifin, A.: A pool based electricity market design for malaysia electricity supply industry. Universiti Teknologi Malaysia (2008)

29. De Pinho, H.: Systems Tools for Complex Health Systems: A Guide to Creating Causal Loop Diagrams. Mailman School Public Health, Columbia University, pp. 1–25 (2015)

30. Coal | 2009–2017 | Data | Chart | Calendar | Forecast | News. In: Trading Econ. https://tradingeconomics.com/commodity/coal. Accessed 8 Aug 2017

31. Clearing the air on tariffs - Nation | The Star Online. Star Online (2017)

32. Najib touts Malaysia's CO2 Successes - Clean Malaysia (2016)

33. TNB: GHG Emissions Reduction Project for Power Sector (2015)

ArUcoRSV: Robot Localisation Using Artificial Marker

Izwan Azmi[1], Mohamad Syazwan Shafei[1],
Mohammad Faidzul Nasrudin[2], Nor Samsiah Sani[2],
and Abdul Hadi Abd Rahman[2(✉)]

[1] Faculty of Information Science and Technology,
Universiti Kebangsaan Malaysia, 43600 Bangi, Selangor, Malaysia
`izwanazmi90@gmail.com`, `syazwan567@gmail.com`
[2] Center for Artificial Intelligence Technology (CAIT),
Faculty of Information Science and Technology, Universiti Kebangsaan
Malaysia, 43600 Bangi, Selangor, Malaysia
`{mfn,norsamsiahsani,abdulhadi}@ukm.edu.my`

Abstract. Robot Soccer (RS) vision system is designed for robot soccer competition. Most RS system implemented colour patches as main marker detector for robot localisation and ball detection. It requires complicated procedures for colour and camera calibration and highly affected by light to run smoothly while it suffers parallax issue. This paper aims to improvise the procedures and performance by minimising the calibration process for robot localisation and produces accurate marker detection for robot identification. ArUcoRSV utilises marker detection using ArUco patches to solve light limitation and able to perform camera calibration processes quick and reliable. It applies an automated perspective using marker detection, refining rejected marker and pose estimation. Experimental results for robot calibration and localisation using ArUcoRSV achieved significant improvements in detection rate with high positional accuracy. This system is expandable and robust to deal with various types of robots including low cost robots.

1 Introduction

Robot Soccer has been the platform for research and development of mobile robot capable of self-reliance involving most areas of engineering and computer science. In general, robot soccer comprises vision system runs on a host computer, which also controls the strategy and communicates to multi robot [1]. Each robot in the system has its own movement mechanism and can move on its own and also work with other robots. Robot Soccer (RS) vision system requires an optimal computational solution especially to perform real-time robot localisation solution and perform desired action. The evaluation of global vision for robot soccer are measured based on 3 factors, (i) frame per second (FPS- the higher the FPS, the better the processing capabilities, (ii) processing speed- the speed taken to process an image and (iii) accuracy- the minimum number of error, the higher the accuracy [2].

© Springer Nature Singapore Pte Ltd. 2019
J.-H. Kim et al. (Eds.): RiTA 2018, CCIS 1015, pp. 189–198, 2019.
https://doi.org/10.1007/978-981-13-7780-8_15

There are several main issues arise in RS vision system development which are camera calibrations and marker detection for robot localisation. Camera calibration is a process to map the position of the robot in the camera image to the physical position of the robot. Pratomo et al. [3] applied neural network transformation real-world coordinates into camera coordinates. Further studies by Pratomo et al. [4] extended to solve camera calibration method for non-linear lens distortion using artificial neural networks. The RS vision system is strongly influenced by the center-offsight system caused by the center-mounted camera. One camera becomes a central image processing system based on individual robots acting autonomously. Therefore, camera calibration is a major process in RS vision to determine accurate robot position for each robot that affected the whole robot operation. Camera calibration in RS vision system is used to change the position of the robot in the camera image to the physical location. It is performed using chessboard calibration grid to determine the exact position of the robot. The data for camera calibration are trained using artificial neural network (ANN). The ANN allows the selection of the camera height and focal length of the lens, then mapping the coordinate transformation from camera coordinates to physical coordinates [4]. After completing the camera calibration process, the computer host sends commands to the robot to start the game.

The second major problem in existing RS is marker detection. Existing robot soccer vision system in Androsot category utilised marker detection using colour detection or colour patch, which makes it difficult for camera calibration [5–7]. Colour calibration is an important process in robot soccer game because it affects colour recognition output in determining robot team members [8, 9]. Each robot is assembled with a colour tag consist of team colour and robot id colour to differentiate according to their roles either goalkeeper, defender or striker. This is to avoid confusion and to detect the error of each team's vision system. In this case, colour space plays an important role in this color determination task. Colour space comes from different colour models such as RGB, YUV and HSV [10]. Colour patch is easily affected in different conditions. Some researchers proposed colour-based solution to solve calibration problem. Li et al. [11] proposed a real-time robust calibration-free colour segmentation method, which can cope with uneven light and lighting condition changes approximate value of color when converting the raw image to HSV color space. Bailey et al. [12] applied an automatic gain control and white balancing within the camera which reduces the effects of the variations in lighting on the thresholds. As a result, the system became more consistent within range of images, significantly simplifying the color segmentation. Apart from detection algorithm, the object needs to be identified. By knowing the colour of the ball and encoding each robot with different colours, it is best to scan the image for the existing pixel within a certain threshold distance of these colours. When a pixel is found, the center of the colour environment is calculated using the center of gravity of all pixels in the locale corresponding to that colour with the threshold. Colour patch is used to determine the robot ID and ball ID in the robot's vision system.

The robot ID means the robot has its own identity. Figure 1 shows a colour patch example used in colour detection techniques. However, colour patches required frequent calibrations when applied in different lighting conditions.

Fig. 1. Colour patch for team and Robot ID (Color figure online)

2 ArUcoRSV

This paper presents an ArUcoRSV to solve camera calibration and robot detection and localisation for various robot heights. An ArUco-based Robot Soccer Vision (Aruco-RSV) system is designed for implementation on a mobile field platform, and expandable with full size robot soccer field. A mobile RS platform is designed for a smaller and more portable scale by is 120 cm (length) × 90 cm (width) × 150 cm (height). It consists of aluminium frame, a wood base covered by a thin black carpet and a stand to hold a Logitech C920 HD Pro Webcam. ArUcoRSV utilizes the using an ArUco ID bookmarking system use of for robot detection and localisation. ArUco marker as in Fig. 2(a) is a rectangular synthetic marker comprising extensive black boundaries and internal binary matrices that define the identifier (ID). Black borders facilitate fast tracking in images and binaries encoding allowing identification and implementation of error tracking and error detection techniques. The marker size determines the size of the internal matrix [13]. ArUco suggests a method for creating a dictionary with a number of markers and the number of bits can be configured. This method maximizes the transitions of bits and inter marker differences in order to reduce the positive error and the inter value error rate respectively. The ArUco library also features a method for error correction. In this study, the ArUco patches size were set to 4 cm × 4 cm.

The tracking process applied an adjustment threshold in a gray-scale image and then searching for a marker candidate by removing the contour, which recognized as rectangle. Furthermore, the code for extraction, identification of the marker and error correction level is executed [13]. ArUco has its own calibration board, which has advantages over OpenCV chessboards. Since the calibration board has some markers, vision is not needed entirely to obtain a calibration point. On the other hand, if the vision is half-calibrated, it was able to execute properly. This advantage is highly important and useful in many applications. Figure 2(b) shows the ArUco calibration board diagram.

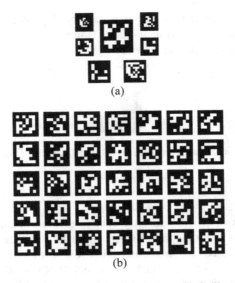

(a)

(b)

Fig. 2. Artificial Marker (a) ArUco Markers (b) Calibration board

2.1 ArUcoRSV System

ArUcoRSV system design involved six interconnect modules; camera, perspective, ArUco, Robot, Ball and Strategy. The relation between each module is shown in Fig. 3. The camera module for providing global vision is used for creating perspective, marker detection using ArUco and Ball detection using color extraction. The perspective and ArUco modules keep updating as the system runs to provide reliable real-time application. Then, the module passed the output of robot localizations to robot module to integrate with ball module for strategy planning.

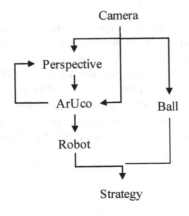

Fig. 3. ArUcoRSV architecture

The flow process for ArUcoRSV program is described in Algorithm 1. There are six main functions; auto brightness calibration, sense, ball tracker, perspective, decide and action. Six on-field points are used to determine the mean value for configuring brightness configuration of the images. This program utilises multi-core programming to synchronise the execution of all six multi-functions in real time implementation.

Algorithm 1. ArUcoRSV program

```
program ArUcoRSV (Output)

begin
  auto brightness calibration:
     6-points brightness;
  sense:
     ArUco marker detection;
     regenerate coordinated;
     geometric image;
  ball tracker:
     color segmentation;
     kalman tracker;
  perspective:
     pixel -> cm
     strategy;
  decide:
     robot id;
  action:
     robot movement;
end.
```

2.2 Automatic Perspective

ArUcoRSV introduces an automatic perspective algorithm to minimise calibration processes. It involves three main processes, which are marker detection; refine detected marker and pose estimation. In marker detection process, system will calculate the contour for each marker will determine the real and rejected markers by utilising marker size, length and width as illustrated in Fig. 4. Then, rejected markers below threshold are refined. The next process is performing pose estimation using pitch, yaw, roll and detected four corners of markers (A, B, C, D) as stated in Eq. 1. The pose estimation also considers camera metric and distortion effects.

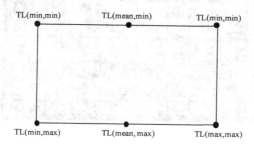

Fig. 4. Automatic perspective using 6-points on field calibration

$$C_{x,y} = (A_{x,y} + B_{x,y} + C_{x,y} + D_{x,y})/4. \tag{1}$$

3   Results and Discussion

The system is developed in the form of standalone software in the Linux operating system. This system uses C++ programming language and uses open source reference library OpenCV version 3.2. This system is developed using Atom IDE software. An 11 × 15 calibration board is used to collect data for robot localisation evaluations. A total of 20539 ArUco markers were collected for three different light intensities in three levels of heights.

3.1   Automatic Perspective

Figure 5(a) shows a field view of the camera. The field looks convex before calibration process using the automatic perspective. After calibration, ArUcoRSV transform the

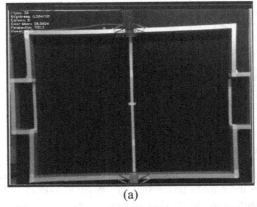

(a)

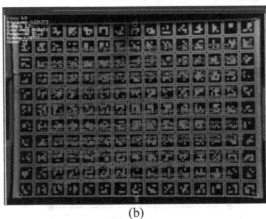

(b)

Fig. 5. The field view (a) before the calibration process (b) using Aruco calibration board.

field based on 6 points field calibrations as in Fig. 5(b). This one-time calibration process is performed automatically without having to change/set any parameters. It performs auto-brightness refining to operate in different lighting conditions.

3.2 Light Intensity

Once the calibration process completed, a series of tests is performed to measure the performance of the system. Table 1 shows the experimental results for light intensity and frame rate in all conditions. The system is able to maintain a significant amount of frame rate range above 15 fps even in low lighting condition and different heights. It shows that light condition doesn't affect much on the marker detection speed, which is important to detect robot movements.

Table 1. Light intensity and frame rate

Light intensity/height	h_0, ground	h_1, 7 cm	h_2, 14 cm	Frame rate
High	100 lx	87 lx	85 lx	15.9 fps
Medium	67 lx	63 lx	61 lx	15.5 fps
Low	5 lx	3 lx	4 lx	15.1 fps

3.3 Detection Rate and Accuracy

ArUcoRSV recorded a 93% of detection rate for all condition and light intensity as tabulated in Table 2. The system produces a high performance for h0 and h1 (95%) for all conditions compare to h2 (90%).

Table 2. Detection rate in different height and light intensity

Height	Light intensity	Total detected marker	Detection rate
h0	High	160	97.10%
	Med	159	96.10%
	Low	152	92.20%
h1	High	149	90.40%
	Med	159	96.20%
	Low	161	97.30%
h2	High	144	87.20%
	Med	151	91.50%
	Low	152	92.20%
Average		152	93.0%

Furthermore, the system detection accuracies are measured in two conditions; based on height and light intensity. Figure 6 shows that the system is highly reliable in low height as for h0 and h1. Detection performances in h2 recorded lower but still considerable. The overall RMSE performance recorded at 29.3 ± 5.6 mm.

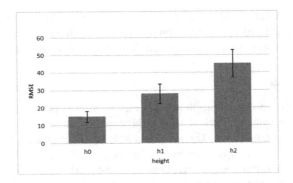

Fig. 6. Detection accuracy in different heights

Next, the detection performances are evaluated based on light intensity; low, medium and high. Figure 7 shows the result of RMSE for all tested light conditions. It is observed that the best accuracy is recorded in medium light condition (65.1 ± 12.2 mm), while still reliable for low and high conditions.

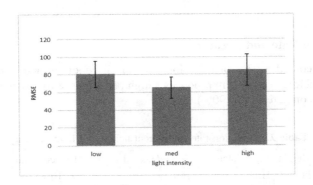

Fig. 7. Detection accuracy in different light intensity

ArUcoRSV produces significant localisation accuracies and capable of dealing with different lighting conditions compare to existing system [3]. It shows that the auto brightness and perspective performs well to overcome these limitations. The positional accuracies achieved for this system in different height with different light intensity means the system successfully deals with parallax issues. The detection rate for multi-points on field calibration shows that ArUco implementation is reliable and suitable for real-time robot soccer application.

4 Conclusion

This paper studies the impact of ArUco marker for robot localization in robot soccer application. The results showed that the camera calibration process using the ArUco marker was easier and flexible compared to color patch techniques while maintaining the accuracies in various lighting and heights conditions. However, several system limitations during where users need to provide the exact ArUco type and marker size parallel to the height of the camera to be able to work successfully. Further work could involve the use of multi-cameras instead of single camera.

Acknowledgment. The authors would like to Universiti Kebangsaan Malaysia, grant ID: KRA-2018-007 for the funding and support for this project.

References

1. Lee, J., Chun, H., Dilmurod, Y., Ko, K.: Robot Intelligence Technology and Applications 2012. Advances in Intelligent Systems and Computing, vol. 208. Springer, Heidelberg (2013). https://doi.org/10.1007/978-3-642-37374-9
2. Nadarajah, S., Sundaraj, K.: Vision in robot soccer: a review. Artif. Intell. Rev. **44**(3), 289–310 (2015)
3. Pratomo, A.H., Zakaria, M.S., Prabuwono, A.S., Liong, C.-Y.: Camera calibration: transformation real-world coordinates into camera coordinates using neural network. In: Omar, K., et al. (eds.) FIRA 2013. CCIS, vol. 376, pp. 345–360. Springer, Heidelberg (2013). https://doi.org/10.1007/978-3-642-40409-2_30
4. Pratomo, A.H., Zakaria, M.S., Nasrudin, M.F., Prabuwono, A.S., Liong, C.-Y., Azmi, I.: Robust camera calibration for the MiroSot and the AndroSot vision systems using artificial neural networks. In: Kim, J.-H., Yang, W., Jo, J., Sincak, P., Myung, H. (eds.) Robot Intelligence Technology and Applications 3. AISC, vol. 345, pp. 571–585. Springer, Cham (2015). https://doi.org/10.1007/978-3-319-16841-8_51
5. Maher, M.M.: Robot Detection Using Gradient and Color Signatures (2016)
6. Chondro, P., Ruan, S.J.: An adaptive background estimation for real-time object localization on a color-coded environment. In: 2016 International Conference on Advanced Computer Science and Information Systems, ICACSIS 2016, pp. 464–469 (2017)
7. Lee, S., Tewolde, G.S., Lim, J., Kwon, J.: Vision based localization for multiple mobile robots using low-cost vision sensor. In: IEEE International Conference on Intelligent Robots and Systems Electro/Information Technology, pp. 280–285 (2015)
8. DeGol, J., Bretl, T., Hoiem, D.: ChromaTag: a colored marker and fast detection algorithm, pp. 1472–1481 (2017)
9. Farazi, H., Allgeuer, P., Behnke, S.: A monocular vision system for playing soccer in low color information environments. In: 10th Workshop on Humanoid Soccer Robot. IEEE-RAS International Conference Humanoid Robots, no. November (2015)
10. Bin Zabawi, N.H., Omar, K.: Robot soccer vision: an overview for new learner. In: 2011 International Conference on Pattern Analysis and Intelligence Robotics, pp. 125–130 (2011)
11. Li, S., Chen, Q.: A real-time robust calibration free color segmentation method for soccer robots

12. Bailey, D., Contreras, M., Sen Gupta, G.: Towards automatic colour segmentation for robot soccer. In: ICARA 2015 – Proceedings of 2015 6th International Conference on Automation, Robotics and Application, pp. 478–483 (2015)
13. Babinec, A., Jurišica, L., Hubinský, P., Duchoň, F.: Visual localization of mobile robot using artificial markers. Proc. Eng. **96**, 1–9 (2014)

Spike Encoding Modules Using Neuron Model in Neural Networks

Yeeun Kim[1(✉)], Seunghee Lee[2], Wonho Song[2], and Hyun Myung[1,2]

[1] Robotics Program, Korean Advanced Institute for Science
and Technology (KAIST), Daejeon 34141, Korea
{yeeunk, hmyung}@kaist.ac.kr
[2] Department of Civil and Environmental Engineering,
Korean Advanced Institute for Science and Technology (KAIST),
Daejeon 34141, Korea
{seunghee.lee, swh4613}@kaist.ac.kr
http://urobot.kaist.ac.kr

Abstract. There has been a great increase in performance of deep neural networks. However, for mobile devices which are not equipped with GPU (Graphics Processing Unit) or powerful CPU (Central Processing Unit), it is still impossible to deal with such a large amount of data in real time. In this paper, preliminary results in spike neural encoding methods reducing the amount of the input and computational load by mimicking the neuronal firing are presented. For this, two neuron models, leaky integrate-and-fire (LIF) model and simplified IF model, are exploited for transforming the input image to the spike image. For the evaluation, MNIST datasets are encoded and tested in deep neural networks for checking the loss of information. The proposed spike encoding modules using neuron models will be able to greatly help reduce required computation by using spike input data in low powered mobile devices.

Keywords: Low-powered · Neural encoding · Spike input · Spike conversion · Spike encoding using neuron model

1 Introduction

Recently, deep neural networks have been surprisingly adopted in various utilization, such as a computer vision [1], language processing [2], sound recognition [3] and so on. Using a huge amount of dataset, there is a growing number of needs for the efficient and considerably accurate neural network. The lighter neural network is considered more significant as the mobile devices and low powered devices, such as cellular phones, tablets, and small sensors are near us. As a part of neural networks, using different kinds of encoding modules could help the entire neural network in reducing the storage and computation [4].

The spiking neural network (SNN) is a neural network using the 'spikes' signaling mechanism propagation. Compared to electrical devices, low energy is consumed in the human brain with outstanding performance in various tasks. The efficiency and performance improvements are expected when the mechanism based on the spike in the

© Springer Nature Singapore Pte Ltd. 2019
J.-H. Kim et al. (Eds.): RiTA 2018, CCIS 1015, pp. 199–206, 2019.
https://doi.org/10.1007/978-981-13-7780-8_16

brain are applied to neural networks [5–10]. Currently, the SNN research is still in the primitive level only having a good performance in customized SNN hardware. Also, spike conversion of input data and spike output data conversion to human comprehensible form are required for utilizing the SNN.

To convert the raw data to the trains of spikes, there are three main techniques [27]: Poisson spike generation [28, 29], rank order encoding [30] and using the output of a silicon retina [31]. Poisson spike generation method encodes the input to the spikes by using Poisson distribution, in this method, if firing rate which is proportional to each pixel value is greater than randomly created value, a spike will be generated. Although the Poisson spike generation is simple to compute, it is not biologically plausible in that it cannot describe the refractory period. Rank Order Encoding is the way that only the order of spikes is stored instead of exact timing. Lastly, the Dynamic Vision Sensor (DVS) can be used for spike generation. However, an additional sensor (DVS) should be implemented for this approach.

This paper introduces the preliminary result of the spike conversion from images using two bio-mimic neural encoding methods, leaky integrate-and-fire (LIF) model and simplified IF model. To implement the two different encoding modules, we first analyzed two neuron models especially on the functions in Sect. 2. Section 3 describes how the spike encoding modules are designed related to the actual spike mechanism, and the experiments and results are explained in Sect. 4. Then, the conclusion and the future works are presented in Sect. 5.

2 Neuron Models

A spiking neuron model describes the properties of the nervous system generating acute electrical potentials across their cell membrane. It is known that spiking neurons are a main signaling unit in the nervous system. There can be two types of neuron models: Electrical input-output membrane voltage models and pharmacological input neuron models. Electrical input-output membrane voltage models predict membrane output voltage when the input current is applied, while pharmacological input neuron models connect the relationship between external stimuli and the probability of a spike occurrence. Here, we will discuss two neuron models of electrical input-output membrane voltage models.

2.1 Leaky Integrate-and-Fire Neuron Model

The leaky integrate-and-fire (LIF) neuron model [11, 12] describes the relationship between input and output membrane voltage. The LIF model is described in Eq. (1)

$$C \cdot \frac{dV_i(t)}{dt} = -g \cdot (V_i(t) - V') - k \cdot A_i(t) \tag{1}$$

where $V_i(t)$ is inner neuron voltage, V' is resting potential voltage (-70 mV), $A_i(t)$ is the external input at time t. C, g, and k denote the membrane capacitance, the passive conductance and the resting membrane potential respectively. As input current is applied, the membrane voltage gets increase by the time reaching threshold voltage. At this time, a spike occurs, causing the voltage to resting potential. Thus, the firing frequency is also increased according to the increase of input current. The weakness of this model is that it does not implement time-dependent memory, however, it is a well-known neuron model.

2.2 Simplified IF Neuron Model

The simplified IF neuron model [13] combines Hodgkin-Huxley [14–17] and IF neuron model for computation effectiveness and biophysically accuracy. The model is described in Eqs. (2) and (3)

$$\frac{dV}{dt} = 0.04\,V^2 + 5\,V + 140 - u + A \tag{2}$$

$$\frac{du}{dt} = a(b\,V - u) \tag{3}$$

$$\text{if } v \geq 30\,\text{mV, then } \begin{cases} v \leftarrow c \\ u \leftarrow u + d \end{cases} \tag{4}$$

where V is the membrane potential of the neuron and u is a membrane recovery variable. A is the external where V is the membrane potential of the neuron and u is a membrane recovery variable. A is the external input. The parameter a is the timescale of u, and b is the sensitivity of u to the subthreshold fluctuations of v. The parameter c is the after-spike reset value of v, and d is the after-spike reset value of u. Equation (2) fits the dynamics of the spike setout in a cortical neuron. The resting potential is set in -70 mV to -60 mV according to the parameter b. The threshold voltage is between -55 mV and -40 mV as the threshold voltage is not fixed like real neurons.

3 Spike Encoding Module

In this section, we give a description of our proposed methods for encoding the input images. For this, we use two different neuron models: leaky integrate-and-fire model and simplified IF model. Firstly, we will introduce how we preprocess the input images to the encoding module. Then we will explain the procedures to convert an image to spike data by using our two encoding modules.

3.1 Image Preprocessing

Before applying the input images to neuron model and converting them to spike data directly, we divided the input images with the $m \times m$ window based on the fact that the real optic nerve accepts the image by dividing it into several parts instead of accepting the whole image at once. Then, to transform the input into the proper form for the neural encoding modules, we linearized the $m \times m$ pixels extracted by the window. As a result, when the size of the input image is $(m \times n) \times (m \times n)$, we obtain $n \times n$ linearized input row vectors (Fig. 1).

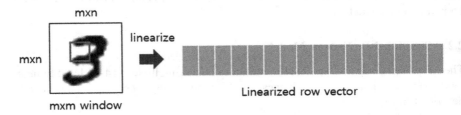

Fig. 1. The process of dividing an input image by window and transforming the pixels to linearized row vectors

3.2   Encoding the Images by Using Neuron Models

Our encoding modules are aimed to convert the input images into spike images which have only two values (fire or not) while the original input images can have 256 different pixel values (0–255) so that the dataset and computation cost get light. For this, firstly we covert the input images composed of pixels into membrane potentials by using the neuron models which illustrate the neuronal firing. Then, if the membrane potential exceeds the threshold we set, we consider that the neuron has fired. Otherwise the neuron doesn't fire. The neuron models that are chosen for calculating the membrane potential are *leaky integrate-and-fire model* [11, 12] and *simplified IF neuron model* [13]. We designed our encoding modules based on these two neuron models.

Encoding Module Using Leaky Integrate-and-Fire Model. To get membrane potential $V_i(t)$, at first, we normalize each value of the linearized vector to be on the interval [0, 1] by dividing by 255 and subtracting them from 1. Then, the linearized vector is put into $A_i(t)$ in Eq. (1). By solving this differential equation with Runge-Kutta method [18], the membrane potentials can be obtained at intervals of 1 ms.

Encoding Module Using Simplified IF Neuron Model. As the linearized vector is normalized in the previous section, in the encoding module using simplified IF neuron model, each value of the linearized vector is normalized to have the value in a range of −5 to 5. For solving the differential equation and obtaining membrane potential, the Euler's method which is a first-order numerical procedure for solving ordinary differential equations is used.

4   Experiment and Results

For validating our encoding modules, we have tested spiked MNIST datasets [19] obtained from our two encoding modules with AlexNet [20] and GoogLeNet [21] and compared their results with the MNIST which consists of handwritten digits and is a total 50,000 training images and 10,000 test image. Figure 2 shows some examples of

MNIST and spiked MNIST datasets. The training and testing of the network were conducted on NVIDIA DIGITS server [22].

In both encoding modules, we simulate each neuron for 100 ms and assign the threshold voltage = −55 mV. If the neuron has fired, we change the value of the membrane potential to 40 mV. To mimic the refractory period of the neuron after firing, we set to the potential voltage −70 mV for 2 ms. So, the membrane voltage is constrained to have the value in a range of −70 mV to 40 mV. And with 4 × 4 window, the 49 × 100 inputs are obtained per an image (28 × 28). Finally, we get spiked images by drawing the lines only if the membrane potential is 40 mV that means the neuron has fired.

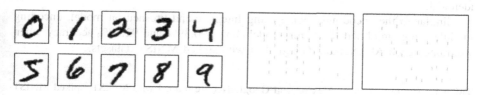

Fig. 2. Examples of MNIST (left) and spiked MNIST (right) datasets.

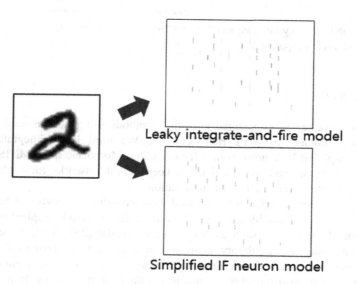

Fig. 3. Converted spiked images by using encoding modules using leaky integrate-and-fire model (above) and simplified IF neuron model (below).

4.1 Spike Encoding Module Using Leaky Integrate-and-Fire Neuron Model

In this experiment, C, g, and k of Eq. (1) are set to 0.5, 15.0 and −1.0, respectively. The examples of converted spiked images are shown in Fig. 3.

As we can see in Table 1 below, when AlexNet and GoogLeNet are trained with spiked MNIST datasets obtained from encoding module using leaky integrate-and-fire neuron model, each network shows 90.76% and 92.86% accuracy, having lower performance compared to original MNIST dataset.

4.2   Spike Encoding Module Using Simplified IF Neuron Model

For testing the performance, we use an encoding module with $a = 0.02$, $b = 0.2$, $c = -70$ and $d = 2$ in Eq. (2–4). The output image by using the encoding module using simplified IF neuron model is also found in Fig. 3. We can see that the outputs are quite different depending on which encoding module is used, even if the input images are identical.

Similar to the encoding module using Integrate-and-fire neuron model, encoding module using simplified IF neuron model shows 90.39% and 92.60% accuracy when we trained each AlexNet and GoogLeNet with spiked MNIST datasets.

Table 1. The results when AlexNet and GoogLeNet are trained with MNIST, spiked MNIST (leaky integrate-and fire) and spiked MNIST (simplified IF).

Dataset	AlexNet (accuracy)	GoogLeNet (accuracy)
MNIST	99.49%	99.45%
spiked MNIST (leaky integrate-and-fire)	90.76%	92.86%
spiked MNIST (simplified IF)	90.39%	92.60%

5   Conclusion

This paper presented spike-inspired encoding methods using neuron models, which reproduced biological neuron's rich behavior. We used leaky integrate-and-Fire (LIF) and simplified IF neuron models for encoding. Moreover, spike-MNIST images encoded by each method are tested using deep neural networks for evaluating the preliminary results about the loss of information.

We expect more accurate and optimized spike-encoding methods can be obtained as further research is done. More diverse neuron models need to be applied for a spike-inspired encoding module, such as Nagumo-Sato model [23]. Also, the concept of excitatory neurons and inhibitory neurons can achieve more biologically-inspired encoding module because it represents that different neurons have different dynamics. In addition, we can properly set the parameters of the neuron model by deep learning as well.

Acknowledgement. This work was supported by the ICT R&D program of MSIP/IITP (2016-0-00563, Research on Adaptive Machine Learning Technology Development for Intelligent Autonomous Digital Companion). The students are supported by Korea Minister of Ministry of Land, Infrastructure and Transport (MOLIT) as U-City Master and Doctor Course Grant Program.

References

1. Krizhevsky, A., Sutskever, I., Hinton, G.: ImageNet classification with deep convolutional neural networks. In: Proceedings of Neural Information Processing Systems (NIPS), pp. 1097–1105 (2012)
2. Hinton, G., et al.: Deep neural networks for acoustic modeling in speech recognition. IEEE Sig. Process. Mag. **29**(6), 82–97 (2012)
3. Mikolov, T., Karafiat, M., Burget, L., Cernocky, J., Khudanpur, S.: Recurrent neural network based language model. In: Interspeech, vol. 2, p. 3 (2010)
4. Han, S., Pool, J., Tran, J., Dally, W.: Learning both weights and connections for efficient neural network. In: Proceedings of Neural Information Processing Systems (NIPS) (2015)
5. Tavanaei, A., Maida, A.: Bio-inspired spiking convolutional neural network using layer-wise sparse coding and STDP learning. arXiv preprint arXiv:1611.03000 (2016)
6. Trabelsi, C., et al.: Deep complex networks. arXiv preprint arXiv:1705.09792 (2017)
7. Courbariaux, M., Hubara, I., Soudry, D., El-Yaniv, R., Bengio, Y.: Binarized neural networks: training deep neural networks with weights and activations constrained to +1 or −1. arXiv preprint arXiv:1602.02830 (2016)
8. Long, L.: Scalable biologically inspired neural networks with spike time based learning. In: Proceedings Learning and Adaptive Behaviors for Robotic Systems, LAB-RS 2008, pp. 29–34. ECSIS (2008)
9. Lee, S., Kim, K., Kim, J., Kim, Y., Myung, H.: Spike-inspired deep neural network design using binary weight. In: Proceedings of International Conference on Control, Automation and Systems (ICCAS) (2018)
10. Kim, K., Kim, J., Lee, S., Kim, Y., Myung, H.: Development of the image to spike conversion algorithm for the deep neural networks. In: Proceedings of International Conference on Control, Automation and Systems (ICCAS) (2018)
11. Abbott, L.F.: Lapicque's introduction of the integrate-and-fire model neuron (1907). Brain Res. Bull. **50**(5–6), 303–304 (1999)
12. Xie, X., Qu, H., Yi, Z., Kurths, J.: Efficient training of supervised spiking neural network via accurate synaptic-efficiency adjustment method. IEEE Trans. Neural Netw. Learn. Syst. **28** (6), 1411–14247 (2017)
13. Izhikevich, E.M.: Simple model of spiking neurons. IEEE Trans. Neural Netw. **14**(6), 1569–1572 (2013)
14. Hodgkin, A., Huxley, A.: A quantitative description of membrane current and its application to conduction and excitation in nerve. J. Physiol. **117**(4), 500–544 (1952)
15. Hodgkin, A., Huxley, A., Katz, B.: Measurement of current-voltage relations in the membrane of the giant axon of Loligo. J. Physiol. **116**(4), 424–448 (1952)
16. Hodgkin, A., Huxley, A.: Currents carried by sodium and potassium ions through the membrane of the giant axon of Loligo. J. Physiol. **116**(4), 449–472 (1952)
17. Hodgkin, A., Huxley, A.: The components of membrane conductance in the giant axon of Loligo. J. Physiol. **116**(4), 473–496 (1952)
18. Runge, C.: Über die numerische Auflösung von Differentialgleichungen. Mathematische Annalen **46**(2), 167–178 (1895). Springer
19. LeCun, Y., Cortes, C., Burges, C.: MNIST handwritten digit database. AT&T Labs, (2018). http://yann.lecun.com/exdb/mnist
20. Krizhevsky, A., Sutskever, I., Hinton, G.: ImageNet classification with deep convolutional neural networks. In: Proceedings of the 25th International Conference on Neural Information Processing Systems (NIPS), pp. 1097–1105 (2012)

21. Szegedy, C., et al.: Going deeper with convolutions. In: Proceedings of the IEEE Conference on Computer Vision and Pattern Recognition (CVPR), pp. 1–9 (2015)
22. NVIDIA: DIGITS Deep Learning Framework (2018). https://github.com/NVIDIA/DIGITS
23. Nemoto, I., Saito, K.: A complex-valued version of Nagumo-Sato model of a single neuron and its behavior. Neural Netw. 15(7), 833–853 (2002)
24. Iakymchuk, T., Rosado-Muñoz, A., Guerrero-Martínez, J., Bataller-Mompeán, M., Francés-Víllora, J.: Simplified spiking neural network architecture and STDP learning algorithm applied to image classification. EURASIP J. Image Video Process. (2015)
25. Bologna, L., et al.: A closed-loop neurobotic system for fine touch sensing. J. Neural Eng. 10 (4), 046019 (2016)
26. Thorpe, S., Delorme, A., Van Rullen, R.: Spike-based strategies for rapid processing. Neural Netw. 14(6-7), 715–725 (2001)
27. Liu, Q., et al.: Benchmarking spike-based visual recognition: a dataset and evaluation. Front. Neurosci. 10, 496 (2016)
28. Heeger, D.: Poisson model of spike generation. Handout Univ. Standford 5, 1–13 (2000)
29. Fatahi, M., et al.: evt_MNIST: a spike based version of traditional MNIST. arXiv preprint arXiv:1604.06751 (2016)
30. Sen, B., Steve, F.: Evaluating rank-order code performance using a biologically-derived retinal model. In: International Joint Conference on IEEE (IJCNN), pp. 2867–2874 (2009)
31. Serrano-Gotarredona, T., Linares-Barranco, B.: A 128 × 128 1.5% contrast sensitivity 0.9% FPN 3 μs latency 4 mW asynchronous frame-free dynamic vision sensor using transimpedance preamplifiers. J. Solid-State Circ. 48(3), 827–838 (2013)

The Classification of EEG Signal Using Different Machine Learning Techniques for BCI Application

Mamunur Rashid[(⊠)], Norizam Sulaiman, Mahfuzah Mustafa,
Sabira Khatun, and Bifta Sama Bari

Faculty of Electrical and Electronics Engineering, Universiti Malaysia Pahang,
26600 Pekan, Pahang, Malaysia
mamun110218@gmail.com,
{norizam,mahfuzah,sabirakhatun}@ump.edu.my,
biftasama_120@yahoo.com

Abstract. Brain-Computer Interface (BCI) or Human-Machine Interface now becoming vital biomedical engineering and technology field which applying EEG technologies to provide assistive device technology (AT) to humans. Hence, this paper presents the results of analyzing EEG signals from various human cognitive states to extract the suitable EEG features that can be employed to control BCI devices which can be used by disabled or paralyzed people. The EEG features in term of power spectral density, spectral centroids, standard deviation and entropy are selected and investigated from two different mental exercises; (i) quick solving math and (ii) relax (do nothing). The selected features then are classified using Linear Discriminant Analysis (LDA), Support Vector Machine (SVM) and K-Nearest Neighbors (k-NN) classifier. Among all these features, the best accuracy have been achieved by the power spectral density. The accuracies of this feature are 95%, 100%, 100% with LDA, SVM and K-NN respectively. Finally, the translation algorithm will be constructed using selected and classified EEG features to control the BCI devices.

1 Introduction

Brain-computer interface (BCI) connects the brain and computer, which enables disabled person to handle different electronic devices with the help of natural impulse. The whole process can be performed without any involvement of human touch. Easy operations of essential devices by disabled people can be ensured by this system. This system is especially suitable for those, who have no control of their normal muscular body to operate the peripheral devices. Moreover, due to user friendly and low cost, this technology is getting popularity day by day. There are a lots of medical and non-medical BCI application for example playing games [1, 2], BCI speller [3], cursor control [4], social interactions by detecting emotions [5], robotic arm control [6],

© Springer Nature Singapore Pte Ltd. 2019
J.-H. Kim et al. (Eds.): RiTA 2018, CCIS 1015, pp. 207–221, 2019.
https://doi.org/10.1007/978-981-13-7780-8_17

wheelchair control [7], home appliances control [8] or smart phone operation using Electroencephalogram (EEG) signals [9]. In emotion recognition system in regard to EEG, emotion is identified and categorized from EEG when the exact stimuli are applied. Because when emotion is changed, the EEG for that emotion is also changed. A plenty of emotion recognition research relating to emotion recognition have been carried out by the BCI researcher in last 20 years. Chakladar, et al. [10] introduced a Correlation-based subset selection technique for dimension lessening as well as used higher order statistical features (mean, skewness, kurtosis, etc.) for classification. They classify four classes of emotion with LDA with the accuracy of 82%. Ahn et al. [11] proposed emotion identification scheme to identify 2 valence classes and 2 arousal classes, which resulted in a combination of 4 fundamental emotions (happy, sorrowful, angry and relaxed) and the neutral state. The authors affirmed the fractal dimension for feature selection and SVM as a classifier whereas the average accuracy across all subjects was 70.5%. Liu et al. [12] designed movie-induced feelings recognizer using EEG. This framework reached 92.26% accuracy in recognizing neutrality from high-arousal with valence emotions and 86.63% to classify negative from positive emotions. They classified 3 positive emotions and 4 negative emotions with 86.43% and 65.09% accuracy respectively. A new method is investigated [13] to select suitable subject-specific frequency bands instead of using permanent frequency bands for the two categories of emotion recognition. Common spatial pattern and SVM have been employed and six subjects were conducted to validate the method. Hence, 74.17% accuracy was achieved for two classes. The paper in [14], reported three classes of human emotion namely sorrowful, excited and relax recognition in real time using wavelet and Learning Vector Quantization (LVQ) with the accuracy of 72% to 87%. In [15], five emotions have been recognized with maximum classification accuracy of 82.87% and 78.57% utilizing entropy feature on 62 channels and 24 channels respectively. Here, wavelet transform is used for the purpose of signal preprocessing. A group of features namely power, standard deviation, variance and entropy have been classified by utilizing K-NN. Happiness, anger and calm have been categorized in [16]. Here, the fractal dimension feature has been classified utilizing RBF SVM with 60% accuracy. In this paper, four feature namely power spectral density; spectral centroids, standard deviation and energy entropy have been classified using LDA, SVM and K-NN. Two classes of mental exercise of eight subjects have been analyzed here. This paper has been organized in the following sections i.e. Sects. 2 and 3 discusses issues related to methodology, results and discussion respectively; finally, Sect. 4 deals with the conclusion.

2   Methodology

A Brain Computer-Interfaces (BCI) aims to provide an alternative communication system by offering human brain a way to control a machines or devices without involving any muscle movements. Figure 1 describes a BCI framework with all

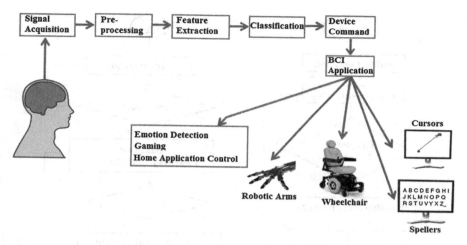

Fig. 1. BCI system with application [17]

elements and applications. A BCI system comes with five elements including signal acquisition, signal preprocessing, feature extraction, classification as well as device command.

Among all these elements, feature selection and classification plays a vital role to design any BCI system because proper feature selection increase the accuracy of the classifier and the BCI device's performance.

BCI researchers have extracted a lots of features. Among them band power spectrum, energy spectral density, spectral centroid, common spatial pattern, wavelet transformations, wavelet packet decomposition, independent component analysis, autoregressive model, principal component analysis, cross-correlation, variant, co-variant, short-time Fourier Transform, Shannon's entropy and z-score are the most usable features [18–21]. These features have been classified by the variety of classification algorithm namely Linear Discriminant Analysis (LDA), Support Vector Machine (SVM), Neural Networks (NN), MultiLayer Perceptron (MLP), Hidden Markov Model (HMM) and K-Nearest Neighbors (K-NN) by the previous research [22]. In this research, average power spectral density, standard deviation, spectral centroid and energy entropy of EEG alpha and beta band have been selected as EEG features and then classified these feature by SVM, K-NN and LDA. Figure 2 represents the flow chart of the methodology.

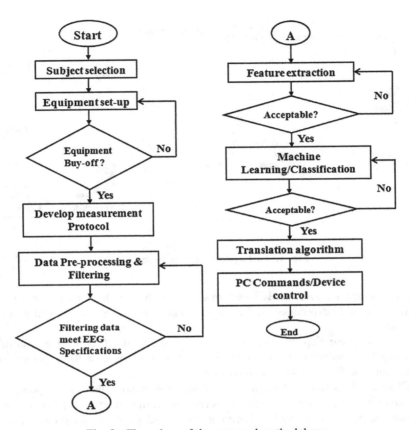

Fig. 2. Flow chart of the proposed methodology.

2.1 Signal Acquisition and Measurement Protocol

The first step of the BCI design is data acquisition. Neurosky Mindwave EEG headset was used for collecting EEG raw data for this research. This EEG headset contains one electrode that is placed on the FP1 area of the human brain. There is a reference electrode which is connected with the ear lobe. This EEG amplifier captures the raw EEG data at 512 Hz sampling rate. The EEG data have been collected from the eight subjects. Table 1 shows all the subjects description.

Table 1. Subjects for EEG data collection.

Subject	Sex	Age	Subject	Sex	Age
Subject-1	Male	25	Subject-5	Male	19
Subject-2	Female	23	Subject-6	Female	19
Subject-3	Male	21	Subject-7	Female	20
Subject-4	Male	22	Subject-8	Male	20

After subject selection, a measurement protocol has been designed (Fig. 3). Here, an android mobile app called eegID in the mobile phone and the Neurosky Mindwave are paired through the Bluetooth.

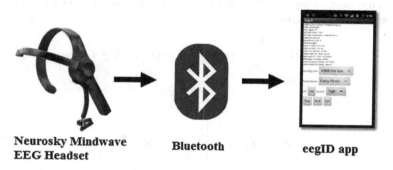

Neurosky Mindwave EEG Headset   **Bluetooth**   **eegID app**

Fig. 3. Raw EEG data acquisition procedure.

Two classes of mental exercise have been considered in this research namely relax (do nothing) and quick math solving. During the data collection of relax state, subjects were said to sit on the chair very comfortably and avoid the movement of whole body. On the other hand, during quick math solving, subjects were said to solve the math problems randomly as quick as possible from the website [23]. Duration of all the data were one minute.

2.2 Data Pre-processing

Generally, there are five frequency bands for each EEG channel, which are delta (0.5–4 Hz), theta (4–8 Hz), alpha (8–13 Hz), beta (14–30 Hz) and gamma (30–45 Hz) [24]. In this research, the EEG data was filtered into alpha and beta band only. The frequency range and activity of EEG alpha and beta band have been shown in Table 2.

Table 2. EEG bands with their activity.

Band	Frequency	Activity
Delta	0.5–4 Hz	Deep sleep
Theta	4–8 Hz	Drowsiness, light sleep
Alpha	8–13 Hz	Relaxed
Beta	13–30 Hz	Active thinking, alert
Gamma	More than 30 Hz	Hyperactivity

2.3 Feature Extraction

Average power spectral density, standard deviation, spectral centroid and energy entropy of EEG alpha and beta band have been analyzed as EEG features in this research.

With the help of Fast Fourier Transformation (FFT), the power spectrum of the EEG data has been measured. The Eq. (1) for FFT [25] and Eq. (2) for power spectral density are shown as follow.

$$X(k) = \sum_{k=1}^{N-1} X(n) W_N^{kn}; \quad K = 0\ldots\ldots N-1 \tag{1}$$

$$W_N = e^{-j\frac{2\pi}{N}}$$

$$PSD = |X(k)|^2 = \left| \sum_{n=0}^{N-1} x(nTs) e^{-j2\pi nk/N} \right|^2 \tag{2}$$

Where one value of 'k' has N complex multiplications, since 'k' = 0, 1… N − 1. The multiplication of x (n) and w^{kn} was done for N times, since n = 0 to N − 1. The Spectral Centroids are measured by utilizing formula in Eq. (3).

$$C = \frac{\int xg(x)dx}{\int g(x)dx} \tag{3}$$

Equation (3) presents the equation of Spectral Centroids that is employed to identify the centre value of the each EEG frequency bands [20].

Conceptually, entropy has to do with how much information is carried by a signal. In other words, entropy can analysis with how much randomness is in the signal. In general, the entropy of a finite length discrete random variable, X = [x(0) x(1) ………x (N − 1)] with probability distribution function denoted by p(x) is defined by

$$H(x) = -\sum_{i=0}^{N-1} P_i(x) \log_2(P_i(x)) \tag{4}$$

Where i represents one of the discrete states. This entropy is larger with the similar probability of occurrence of each discrete state. The LogEn entropy of x is defined by [26],

$$H_{logEn} = -\sum_{i=0}^{N-1} (\log_2(P_i(x)))^2 \tag{5}$$

2.4 Classification

In this research, LDA, SVM and K-NN have been applied to classify the selected features. Each classifier has been employed to every feature as well as the combination of all features to find the batter result.

2.4.1   Linear Discriminant Analysis (LDA)

LDA is deployed to find the linear combinations of feature vectors which describe the characteristics of the corresponding signal. LDA seeks to separate two or more classes of objects or events representing different classes. It utilizes hyperplanes to accomplish this mission. Isolating hyperplane is acquired by looking for the projection which maximizes the distance between the classes' means whereas the interclass variance are minimized. This technique has a very small computational requirement and it is simple to use. Subsequently, LDA has been utilized with success in BCI systems like P300 speller, motor imagery based BCI as well as multiclass BCI [27].

2.4.2   Support Vector Machine (SVM)

SVM is an algorithm that belongs to a category of classification methods which use supervised learning to separate two different classes of data. It utilizes a discriminant hyperplane to detect classes like LDA. However in SVM, the selected hyperplane is the one that maximizes the distance from the closest training points. This optimal hyperplane is represented by the vectors which lie on the margin which are called support vectors, such an SVM enables classification using linear decision boundaries which is designated as linear SVM. This classifier has been employed always with success to a relatively excessive number of synchronous BCI issues [28]. However, it is possible to create nonlinear decision boundaries using a kernel function. The Gaussian or Radial Basis Function (RBF) kernel which are commonly utilized in BCI research is familiarized as Gaussian SVM or RBF SVM [29].

2.4.3   K-Nearest Neighbors (K-NN)

The focus of this approach is to assign an unseen point of the dominant class amongst its k nearest neighbors inside the training set [30]. For BCI designs, these nearest neighbors are generally obtained employing a metric distance. With adequately high value of k and enough training samples, K-NN can approximate any function which empowers it to create nonlinear decision boundaries. The foremost benefit of K-NN algorithm is its simplicity.

3   Result and Discussion

The raw EEG data of subject-1 in time domain and frequency domain have been shown in Fig. 4 both for quick math solving and relax states. After raw data collection, the data have been filtered into alpha and beta band. Figure 5 shows EEG alpha and beta band of subject-1 in time domain. Similarly, Fig. 6 shows EEG alpha and beta band of subject-1 in frequency domain.

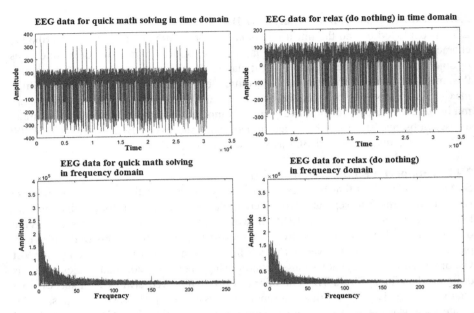

Fig. 4. Plotting of EEG raw data in time and frequency domain for subject-1

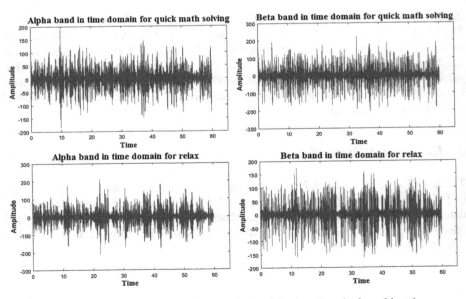

Fig. 5. Plotting of EEG Alpha and Beta band in time domain for subject-1.

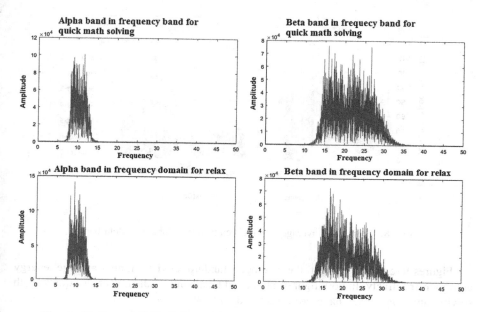

Fig. 6. Plotting of EEG Alpha and Beta band in frequency domain for subject-1.

Different types of EEG features have been analyzed from the filtered EEG alpha and beta band. Figure 7 shows the average power spectral density of alpha and beta band where the beta bands are higher than the alpha band in both mental states. In Fig. 8, the average spectral centroid of alpha and beta for all subject with two mental exercise band have been plotted and here the amplitude of beta bands are also higher than the alpha band.

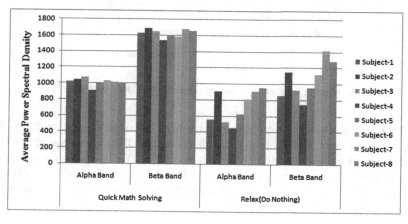

Fig. 7. Plotting of average power spectral density of Alpha and Beta band.

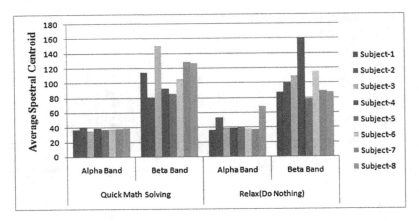

Fig. 8. Plotting of average spectral centroid of Alpha and Beta band.

Figures 9 and 10 present the average standard deviation and average energy entropy respectively. In both figures, alpha and beta band of all subjects for quick math solving and relax states have been analyzed.

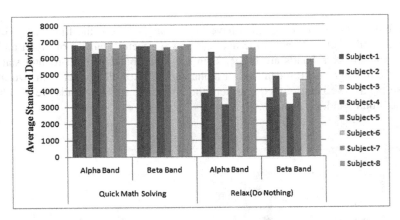

Fig. 9. Plotting of average standard deviation of Alpha and Beta band.

After features selection, every feature have been classified separately. LDA, SVM and K-NN have been applied to classify these feature. Different types of SVM and KNN have been applied to fine out the best technique. Table 3 shows the best classification accuracy of different classifier with selected features. The ratio of training and testing data was 62.5: 37.5. After classifying each feature separately, the combinations of all features have also been classified.

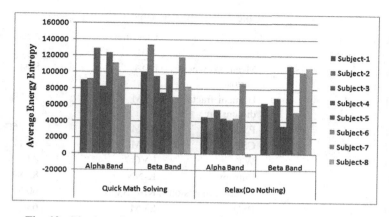

Fig. 10. Plotting of average energy entropy of Alpha and Beta band.

Table 3. Classification accuracy of the selected features.

Feature	Classifier	Classifier accuracy
Power spectral density	LDA	95%
	Cubic SVM	100%
	Fine K-NN	95%
Spectral centroid	LDA	60%
	Quadratic SVM	50%
	Medium K-NN	55%
Standard deviation	LDA	95%
	Cubic SVM	95%
	Fine K-NN	95%
Log energy entropy	LDA	85%
	Linear SVM	75%
	Cubic KNN	65%
Combination of power spectral density, spectral centroid, standard deviation, log energy entropy	LDA	95%
	Cubic SVM	95%
	Weighted K-NN	95%

From Table 3 it is clear that the average power spectral density gives the highest accuracy with all the classification algorithm. The classification accuracy of average power spectral density with LDA, cubic SVM and fine K-NN are 95%, 100% and 100% respectively. The results of proposed method have also been compared with the previous research shown in Table 4.

Table 4. Comparison table

Reference	Class of mental exercise	Features	Classifier	Accuracy
[10]	4	Mean, skewness, kurtosis	LDA	82%
[11]	4	Higuchi fractal dimension	SVM	70.5%
[12]	2	PSD	SVM	92.26%
[13]	2	CSP	SVM	74.17
[14]	3	WT	LVQ	87%
[15]	5	Entropy	K-NN	82.87
[16]	3	Fractal dimension	RBF SVM	60%
[31]	3	PSD	SVM	63%
[32]	2	PSD	SVM	58.8%
[33]	5	Entropy	K-NN	83%
[34]	2	PSD	K-NN	70%
[35]	2	PSD	SVM	85.4%
[36]	5	AR and FFT	SVM	88.51%
[37]	–	Wavelet analysis (RBF)	ANN	84.5%
Proposed method	2	Power spectral density	Cubic SVM	100%
		Spectral centroid	LDA	60%
		Standard deviation	LDA, Cubic SVM, Fine K-NN	95%
		Log energy entropy	LDA	85%

4   Conclusion

In this research, two classes of mental exercise have been classified with the four EEG features namely Power Spectral Density, Spectral Centroid, Standard Deviation and Energy Entropy. To classify these features, LDA, SVM and K-NN classification algorithm have been used here. The best classification accuracy has been obtained with the power spectral density of EEG alpha and beta band with cubic SVM. Thus, the classifier result will be applied to the translation algorithm to run the BCI application such as to control the BCI device like wheelchair.

Acknowledgment. The author would like to acknowledge the great supports by his postgraduate supervisor, research team members, Faculty of Electrical & Electronics Engineering as well as Universiti Malaysia Pahang for providing financial support through research grant, RDU180396.

References

1. van de Laar, B., et al.: Experiencing BCI control in a popular computer game. IEEE Trans. Comput. Intell. AI Games **5**(2), 176–184 (2013)
2. Jiang, D., Yin, J.: Research of auxiliary game platform based on BCI technology. In: Asia-Pacific Conference on Information Processing, APCIP 2009, pp. 424–428 (2009)
3. Vo, K., Nguyen, D.N., Kha, H.H., Dutkiewicz, E.: Real-time analysis on ensemble SVM scores to reduce P300-Speller intensification time. In: 2017 39th Annual International Conference of the IEEE Engineering in Medicine and Biology Society (EMBC), Seogwipo, pp. 4383–4386 (2017)
4. Aydemir, O., Kayikcioglu, T.: Decision tree structure based classification of EEG signals recorded during two dimensional cursor movement imagery. J. Neurosci. Methods **229**, 68–75 (2014). ISSN 0165-0270
5. Zhang, B., Jiang, H., Dong, L.: Classification of EEG signal by WT-CNN model in emotion recognition system. In: 2017 IEEE 16th International Conference on Cognitive Informatics and Cognitive Computing (ICCI*CC), Oxford, pp. 109–114 (2017)
6. Latif, M.Y., et al.: Brain computer interface based robotic arm control. In: 2017 International Smart Cities Conference (ISC2), Wuxi, pp. 1–5 (2017)
7. Singla, R., Khosla, A., Jha, R.: Influence of stimuli colour in SSVEP-based BCI wheelchair control using support vector machines. J. Med. Eng. Technol. **38**(3), 125–134 (2014)
8. Anindya, S.F., Rachmat, H.H., Sutjiredjeki, E.: A prototype of SSVEP-based BCI for home appliances control. In: 2016 1st International Conference on Biomedical Engineering (IBIOMED), Yogyakarta, pp. 1–6 (2016)
9. Kumar, P., Saini, R., Sahu, P.K., Roy, P.P., Dogra, D.P., Balasubramanian, R.: Neurophone: an assistive framework to operate smartphone using EEG signals. In: 2017 IEEE Region 10 Symposium (TENSYMP), Cochin, pp. 1–5 (2017)
10. Chakladar, D.D., Chakraborty, S.: EEG based emotion classification using "correlation based subset selection". Biol. Inspired Cogn. Arch. **24**, 98–106 (2018). ISSN 2212-683X
11. Anh, V.H., Van, M.N., Ha, B.B., Quyet, T.H.: A real-time model based support vector machine for emotion recognition through EEG. In: 2012 International Conference on Control, Automation and Information Sciences (ICCAIS), Ho Chi Minh City, pp. 191–196 (2012)
12. Liu, Y.-J., Yu, M., Zhao, G., Song, J., Ge, Y., Shi, Y.: Real-time movie-induced discrete emotion recognition from EEG Signals. IEEE Trans. Affect. Comput. 1 (2017). https://doi.org/10.1109/taffc.2017.2660485
13. Pan, J., Li, Y., Wang, J.: An EEG-based brain-computer interface for emotion recognition. In: 2016 International Joint Conference on Neural Networks (IJCNN), Vancouver, BC, pp. 2063–2067 (2016)
14. Djamal, E.C., Lodaya, P.: EEG based emotion monitoring using wavelet and learning vector quantization. In: 2017 4th International Conference on Electrical Engineering, Computer Science and Informatics (EECSI), Yogyakarta, pp. 1–6 (2017)
15. Murugappan, M.: Human emotion classification using wavelet transform and KNN. In: 2011 International Conference on Pattern Analysis and Intelligence Robotics, Putrajaya, pp. 148–153 (2011)
16. Kaur, B., Singh, D., Roy, P.P.: EEG based emotion classification mechanism in BCI. Procedia Comput. Sci. **132**, 752–758 (2018). ISSN 1877-0509
17. Ortiz-Rosario, A., Adeli, H.: Brain-computer interface technologies: from signal to action. Rev. Neurosci. **24**(5), 537–552 (2013)

18. Knott, V., Mahoney, C., Kennedy, S., Evans, K.: EEG power, frequency, asymmetry and coherence in male depression. Psychiatry Res.: Neuroimaging **106**(2), 123–140 (2001)
19. Chaouachi, M., Jraidi, I., Frasson, C.: Modeling mental workload using EEG features for intelligent systems. In: Konstan, J.A., Conejo, R., Marzo, J.L., Oliver, N. (eds.) UMAP 2011. LNCS, vol. 6787, pp. 50–61. Springer, Heidelberg (2011). https://doi.org/10.1007/978-3-642-22362-4_5. The cognitive activation theory of stress. Psychoneuroendocrinology **29**, 567–592 (2004)
20. Sulaiman, N., Taib, M.N., Lias, S., Murat, Z.H., Aris, S.A.M., Hamid, N.H.A.: Novel methods for stress features identification using EEG signals. Int. J. Simul. Syst. Sci. Technol. **12**(1), 27–33 (2011)
21. Shen, K.Q., Ong, C.J., Li, X.P., Hui, Z., Wilder-Smith, E.P.V.: A feature selection method for multilevel mental fatigue EEG classification. IEEE Trans. Biomed. Eng. **54**(7), 1231–1237 (2007)
22. Lotte, F., Congedo, M., Lécuyer, A., Lamarche, F., Arnaldi, B.: A review of classification algorithms for EEG-based brain–computer interfaces. J. Neural Eng. **4**, 24 (2007). <inria-00134950>
23. https://arithmetic.zetamac.com/
24. Atkinson, J., Campos, D.: Improving BCI-based emotion recognition by combining EEG feature selection and kernel classifiers. Expert Syst. Appl. **47**, 35–41 (2016)
25. Otsuka, T., et al.: Effects of mandibular deviation on brain activation during clenching: an fMRI preliminary study. Cranio **27**, 88–93 (2009)
26. Aydın, S., Saraoğlu, H.M., Kara, S.: Log energy entropy-based EEG classification with multilayer neural networks in seizure. Ann. Biomed. Eng. **37**(12), 2626–2630 (2009)
27. Cui, G., Zhao, Q., Cao, J., Cichocki, A.: Hybrid-BCI: classification of auditory and visual related potentials. In: 2014 Joint 7th International Conference on Soft Computing and Intelligent Systems (SCIS) and 15th International Symposium on Advanced Intelligent Systems (ISIS), Kitakyushu, pp. 297–300 (2014)
28. Hortal, E., Iáñez, E., Úbeda, A., Planelles, D., Costa, Á., Azorín, J.M.: Selection of the best mental tasks for a SVM-based BCI system. In: 2014 IEEE International Conference on Systems, Man, and Cybernetics (SMC), San Diego, CA, pp. 1483–1488 (2014)
29. Jian, H.L., Tang, K.T.: Improving classification accuracy of SSVEP based BCI using RBF SVM with signal quality evaluation. In: 2014 International Symposium on Intelligent Signal Processing and Communication Systems (ISPACS), Kuching, pp. 302–306 (2014)
30. Bose, R., Khasnobish, A., Bhaduri, S., Tibarewala, D.N.: Performance analysis of left and right lower limb movement classification from EEG. In: 2016 3rd International Conference on Signal Processing and Integrated Networks (SPIN), Noida, pp. 174–179 (2016)
31. Chanel, G., Kierkels, J.J., Soleymani, M., Pun, T.: Short-term emotion assessment in a recall paradigm. Int. J. Hum.-Comput. Stud. **67**, 607–627 (2009)
32. Koelstra, S., et al.: Single trial classification of EEG and peripheral physiological signals for recognition of emotions induced by music videos. In: Yao, Y., Sun, R., Poggio, T., Liu, J., Zhong, N., Huang, J. (eds.) BI 2010. LNCS (LNAI), vol. 6334, pp. 89–100. Springer, Heidelberg (2010). https://doi.org/10.1007/978-3-642-15314-3_9
33. Murugappan, M., Nagarajan, R., Yaacob, S.: Combining spatial filtering and wavelet transform for classifing human emotions using EEG Signals. J. Med. Biol. Eng. **31**, 45–51 (2011)
34. Bastos-Filho, T.F., Ferreira, A., Atencio, A.E., Arjunan, S., Kumar, D.: Evaluation of feature extraction techniques in emotional state recognition. In: 2012 4th International Conference on Intelligent Human Computer Interaction (IHCI), pp. 1–6 (2012)

35. Jatupaiboon, N., Pan-ngum, S., Israsena, P.: Emotion classification using minimal EEG channels and frequency bands. In: 2013 10th International Joint Conference on Computer Science and Software Engineering (JCSSE), pp. 21–24 (2013)
36. Lokannavar, S., Lahane, P., Gangurde, A., Chidre, P.: Emotion recognition using EEG signals. Emotion **4**, 54–56 (2015)
37. Srinivas, V.: Wavelet based emotion recognition using RBF algorithm (2016). https://doi.org/10.17148/IJIREEICE.2016.4507

The Identification of *Oreochromis niloticus* Feeding Behaviour Through the Integration of Photoelectric Sensor and Logistic Regression Classifier

Mohamad Radzi Mohd Sojak, Mohd Azraai Mohd Razman[✉],
Anwar P. P. Abdul Majeed, Rabiu Muazu Musa,
Ahmad Shahrizan Abdul Ghani, and Ismed Iskandar

Innovative Manufacturing, Mechatronics, and Sports Laboratory,
Faculty of Manufacturing Engineering, Universiti Malaysia Pahang,
26600 Pekan, Pahang, Malaysia
azraai@ump.edu.my

Abstract. *Oreochromis niloticus* or tilapia is the second major freshwater aquaculture bred after catfish in Malaysia. By understanding the feeding behaviour, fish farmers will able to identify the best feeding routine. In the present investigation, photoelectric sensors are used to identify the movement, speed and position of the fish. The signals acquired from the sensors are converted into binary data. The hunger behaviour classes are determined through k-means clustering algorithm, i.e., satiated and unsatiated. The Logistic Regression (LR) classifier was employed to classify the aforesaid hunger state. The model was trained by means of 5-fold cross-validation technique. It was shown that the LR model is able to yield a classification accuracy for tested data during the day at three different time windows (4 h each) is 100%, 88.7% and 100%, respectively, whilst the for-night data it was shown to demonstrate 100% classification accuracy.

Keywords: Photoelectric sensor · Logistic regression ·
Oreochromis niloticus · Fish hunger behaviour

1 Introduction

Oreochromis niloticus or tilapia is the second major freshwater aquaculture bred after catfish in Malaysia. It contributed RM 329 million towards the freshwater aquaculture industry in 2013. Tilapia has both fast breeding cycles and weight gain, in addition to its affordable source of protein to the masses. Fish feed management is a non-trivial aspect in aquaculture to meet the ever-increasing food demand and security. More often than not, different fish species has its own unique feeding behaviour. This feeding behaviour is dependent on several factors namely underfeeding, inhibits growth and also competition among their species [1].

Researchers have shown an increased interest in investigating the behaviour of full fed fish (satiated) and less fed fish for better understand its scavenging characteristics in controlled space (tank). It was reported that the fish moves about and appears to be more active during unsatiated state, whilst appears to be dormant if it is satiated [2].

© Springer Nature Singapore Pte Ltd. 2019
J.-H. Kim et al. (Eds.): RiTA 2018, CCIS 1015, pp. 222–228, 2019.
https://doi.org/10.1007/978-981-13-7780-8_18

Research carried out on the feeding character of the Tilapia fish species demonstrated three distinct categories namely; dominants, sub-dominants and subordinates. The dominant fish is the most active fish during feeding as the dominant fish tend to eat more than the others, the sub-dominants and subordinates are characterised with a conflict avoidance behaviour during feeding. Bold fish tend to take more risks and explore their environment, on the other hand, non-dominants and subordinates fish are less aggressive towards food exploration [1].

The effect of lighting conditions towards hunger behaviour has also been investigated [3]. It was shown in the study that exposing the fish under ultradian rhythm of light and dark environment that replicates day and night moment pulses does influence the hunger behaviour. Image processing technique has also been used to aid hunger identification of *Lates Clacarifer* through its integration with different machine learning algorithms, i.e., a variation of *k*-Nearest Neighbour and Support Vector Machine algorithms [4–6]. It was shown from the studies that the application of machine learning is able to identify hunger state of the aforesaid fish species with reasonable classification accuracies.

The present study aims at evaluating the use of photoelectric sensors along with the use of the logistic regression classifier in identifying hunger behaviour of Tilapia. The content of the paper is as follows: Sect. 2 describes the methodology including the experimental setup, the clustering technique and the classification method. Conversely, Sect. 3 details on the results as well as the discussion with regards to the results obtained. Section 4 concludes the present investigation and propose recommendations for future works.

2 Methodology

2.1 Type of Fish

The type of fish used in this study is the *Oreochromis niloticus* (Tilapia) commonly found in Malaysia which they demand food during day and night time. Initial observation suggests that the fish appears to swim slow and be in a sedentary position (usually at the base of the aquarium) after being fed (state of satiated). Before feeding, it is observed that the fish tends to be in a scavenging behaviour (hungry). In this research, 15 juvenile Tilapia are fed once during the daytime. The fish are fed manually, and the feeding is stopped once, the fish food is left to float and untouched by the Tilapia. It is worth noting that uneaten fish food can contaminate and reduce the water quality in the fish tank [7].

2.2 Experiment Setup

A fish tank with a dimension of (36-in. × 18-in. × 19 in.) is used in this investigation. The photoelectric sensors (E3Z-R61Z) denoted as S1, S2, S3, S4, respectively are placed on the right side with a 3D printed holder as shown in Fig. 1. The particular photoelectric sensor model used requires a reflector for each sensor, hence the reflectors are positioned to be aligned with its respective sensor on the left end of the fish tank. Moreover, the sensors are spaced 7.62 cm between each other. The data acquisition module used in this system is Arduino Uno whilst the CooltermWin software is used to collect and save the data. The data is collected continuously for 24 h.

Fig. 1. Experiment setup using four photoelectric sensors in a fish tank.

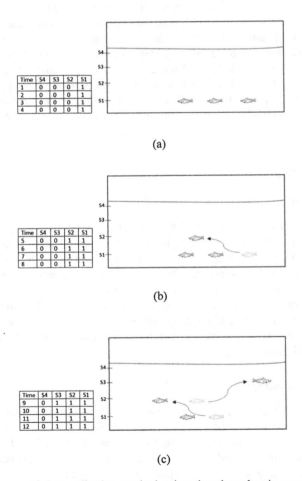

Time	S4	S3	S2	S1
1	0	0	0	1
2	0	0	0	1
3	0	0	0	1
4	0	0	0	1

(a)

Time	S4	S3	S2	S1
5	0	0	1	1
6	0	0	1	1
7	0	0	1	1
8	0	0	1	1

(b)

Time	S4	S3	S2	S1
9	0	1	1	1
10	0	1	1	1
11	0	1	1	1
12	0	1	1	1

(c)

Fig. 2. The process of data collection method using the photoelectric sensor to detect fish position.

As shown in Fig. 2(a) above, the fish is positioned along S1, and it is detected as '1' throughout the time stamp. As one of the fish moved to S2 depicted in Fig. 2(b), the sensor reads '1'. Similarly, as shown in Fig. 2(c), S1, S2, and S3, read '1' as fish are located at different positions. Figure 3 shows the data stream of the position of the fish for a period of time, i.e. 10.15 a.m. to 10.25 a.m. The y-axis informs the position of the fish. Although four sensors are used, nonetheless, seven positions are derived. S1, S2, S3 and S4, corresponds to the fish position at 1, 3, 5 and 7, respectively, whilst positions 2, 4 and 6 are located between S1 and S2, S2 and S3, S3 and S4, respectively. In the event that fish are located at both S1 and S2, then it is assumed that the fish is likely to be located at position 2. If the fish is located between S1 and S3, then the likeliness of the position of the fish will be at position 3.

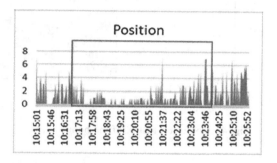

Fig. 3. The position and time stamp data collection of the fish hunger. The Red box shows the changes of position after 20 min of feeding (Color figure online).

2.3   Clustering and Classification

The data obtained from the experimental setup throughout 24 h are clustered by means on *k*-means technique. The features selected to aid the clustering of the data are Position, S1, S2, S3 and S4. The Euclidean distance metric was used to evaluate the data and to cluster the behaviour to unsatiated, C1 and satiated, C2. The clustering was carried out by means of Orange v3.11. The Logistic Regression model developed through MATLAB 2016a is used to classify the clusters identified. A total of 600 data is used to train the model. The five-fold cross-validation technique was used to validating the model as it has been reported to be able to mitigate the issue of overfitting [6, 8]. The classification accuracy metric is used to evaluate the performance of the model. In addition, fresh data was used to further evaluate the efficacy of the model developed.

3   Results and Discussions

Figure 4 illustrates the confusion matrix of the LR model developed on 600 data. It is demonstrated that the model is able to classify the hunger state, i.e. C1 (unsatiated) and C2 (satiated) exceptionally well without no misclassification of the clustered data.

display the clustered data after defining the centres based on each of the data. The confusion matrix shown represents the hunger state of the Tilapia for 30 min out of 24 h collected by the photoelectric sensor during the daytime (Fig. 5).

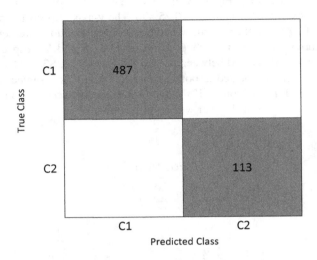

Fig. 4. Confusion matrix of the hunger state.

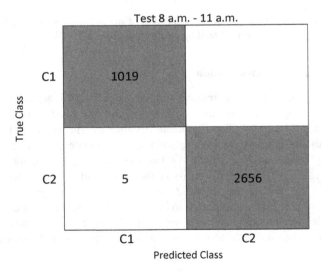

Fig. 5. A sample of the confusion matrix evaluated on fresh data from 8 a.m. to 11 p.m.

Tables 1 and 2 tabulate the classification accuracy for the system in determining the hunger state of fresh data. The data tested consists of 3 h interval for both day and night. It could be seen from the table that the LR model developed is able to reasonably classify accurately the hunger state of the fish. It is apparent that there is no misclassification transpired for the night condition. This observation might be related to the Circadian rhythm of the fish that suggest the fish hunger feeding behaviour is better elucidated at night time instead of the day. Moreover, the reasonably good classification accuracy attained by the LR model could be associated with the selection of the features used in the present investigation.

Table 1. The accuracy of 3 h interval data taken from 8 a.m.–7 p.m.

Day	
Hours	Overall accuracy
8 a.m.–11 a.m.	99.86%
12 a.m.–3 p.m.	84.08%
4 p.m.–7 p.m.	100.00%

Table 2. The accuracy of 3 h interval data taken from 5 p.m.–1.30 a.m.

Night	
Hours	Overall accuracy
5 p.m.–8 p.m.	100.00%
8.20 p.m.–11.20 p.m.	100.00%
10.20 p.m.–1.30 a.m.	100.00%

4 Conclusion

This study evaluated the efficacy of the LR classifier in ascertaining hunger state of Tilapia fish species through the data acquired from photoelectric sensors. It was shown that the experimental setup, which is deemed to be low-cost is non-trivial in aiding the data acquisition. Moreover, the clustering and classification techniques employed yielded desirable results that could probably due to the selection of the features. Therefore, the proposed setup is feasible to be implemented in large-scale aquaculture industry. Future study will look into the efficacy of other machine learning algorithms as well as the integration of fusion sensors in determining the hunger state.

Acknowledgement. The final outcome of this research project and the successful of development of this useful system required a lot of guidance and assistance from my project supervisor. Meanwhile, I would like to express my gratitude to lab instructors and my friends for providing practically knowledge, skills and guidance when doing the mechanical work in lab. This work is partially support by Universiti Malaysia Pahang, Automotive Engineering Centre (AEC) research grant RDU1803131 entitled Development of Multi-vision guided obstacle Avoidance System for Ground Vehicle.

References

1. Benhaïm, D., Akian, D.D., Ramos, M., Ferrari, S., Yao, K., Bégout, M.L.: Self-feeding behaviour and personality traits in tilapia: a comparative study between *Oreochromis niloticus* and *Sarotherodon melanotheron*. Appl. Anim. Behav. Sci. **187**, 85–92 (2017)
2. Hansen, M.J., Schaerf, T.M., Ward, A.J.W.: The effect of hunger on the exploratory behaviour of shoals of mosquitofish *Gambusia holbrooki*. Behaviour **152**, 1659–1677 (2015)
3. Sanchez-Vázquez, F.J., Madrid, J.A., Zamora, S.: Circadian rhythms of feeding activity in sea bass, Dicentrarchus labrax L.: dual phasing capacity of diel demand-feeding pattern. J. Biol. Rhythms **10**, 256–266 (1995)
4. Taha, Z., et al.: The identification of hunger behaviour of *Lates Calcarifer* through the integration of image processing technique and support vector machine. In: IOP Conference of Series of Materials Science and Engineering, vol. 319, p. 012028 (2018)
5. Taha, Z., et al.: The classification of hunger behaviour of *Lates Calcarifer* through the integration of image processing technique and k-Nearest Neighbour learning algorithm. In: IOP Conference of Series of Materials Science and Engineering, vol. 342, p. 012017 (2018)
6. Taha, Z., et al.: The Identification of hunger behaviour of Lates Calcarifer using k-nearest neighbour (2018)
7. Siddiqui, S.A., et al.: Automatic fish species classification in underwater videos: exploiting pre-trained deep neural network models to compensate for limited labelled data. ICES J. Mar. Sci. **75**, 374–389 (2018)
8. Muazu Musa, R., Taha, Z., Abdul Majeed, A.P.P., Abdullah, M.R.: Machine Learning in Sports: Identifying Potential Archers. SAST. Springer, Singapore (2019). https://doi.org/10.1007/978-981-13-2592-2

Motion Planning to Reduce the Thrust of Underwater Robot Using Null-Space Compliance

Junbo Chae, Youngeon Lee, Yeongjun Lee, Hyun-Taek Choi,
and Tae-Kyeong Yeu$^{(\boxtimes)}$

Marine ICT Research Division, KRISO, Daejeon, Korea
{lisser,yglee,leeyeoungjun,htchoi,yeutk}@kriso.re.kr

Abstract. This paper proposes a motion planning method that reduces the thrust of underwater robots moving a manipulator. The proposed method plans motion with 12-degree-of-freedom (12-DOF) for underwater robot, including 6-DOF for Body and 6-DOF for the manipulator. 2-DOF are used for the body's orientation, roll and pitch and 3-DOF for manipulator position and orientation each. While the robot performs a task, if the thrust of the underwater robot is close to the limit of thruster, the underwater robot reduces its thrust by moving the center of the mass (COM) of the manipulator using null-space control without disturbing the task. The proposed method has been verified via simulation, by planning motion such that the thrust of the underwater robot is reduced under the limit of thrusters.

1 Introduction

Maintaining an underwater robot's attitude is critical in autonomous manipulation so as not disturb the position of a manipulator's end-effector. When the manipulator reaches out to grasp an object, the underwater robot's thrusters work to maintain the attitude of the robot body compensating for the weight of the manipulator. The heavy weight of a manipulator can cause problems to maintain attitude, when the thrusters do not have enough thrust.

If the robot has a redundancy of the DOF, the robot can make another motion without disturbing the main task. Zghal et al. avoided the limit joint of joint angle using redundancy [1] and Nakamura and Hanafusa used redundancy for making the robust singularity avoidance algorithm [2]. Antonelli, Chiaverini used redundancy for making the task priority for the underwater robot [3] Kang et al. used redundancy for dynamic stability of underwater using zero moment point [4].

In this paper, a motion planning method decreasing the underwater robot's thrust using null-space compliance is proposed. If the thrust of the robot is higher than the thrust limit, we move the position of the robot body and manipulator joint angle to reduce the force and torque due to the manipulator's weight by moving the position of the center of the mass (COM) of the manipulator. The desired position and orientation of the robot body and the manipulator are provided using null-space so as not disturb

J.-H. Kim et al. (Eds.): RiTA 2018, CCIS 1015, pp. 229–235, 2019.
https://doi.org/10.1007/978-981-13-7780-8_19

the main task. The result of motion planning was performed in a simulator, free-floating gazebo that supports underwater dynamic simulation [5].

2 System Configuration

2.1 Model

In this paper, we constructed a 12-DOF robot including 6-DOF for the body and 6-DOF for the manipulator. The simulation is shown in Fig. 1. The robot was manufactured made by Korea Research Institute of Ship and Ocean Engineering (KRISO). It has eight thrusters, of which there are four vertical thrusters and four horizontal thrusters. These thrusters are used to control the body's position and orientation. The model of the manipulator used is the ARM 7E mini of the Eca group having six joints and one gripper.

Fig. 1. 12-DOF underwater robot with Arm 7E mini manipulator in simulation.

The kinematics of the robot body and the manipulator are shown in Fig. 2. The robot body has six virtual joints that comprises three prismatic joints that control the position of the robot body and three revolute joints that control the orientation of the robot body. q_1, q_2, q_3 denote the virtual prismatic joints of the robot body for the z, y, x Cartesian coordinates, respectively, in the global frame. q_4, q_5, q_6 denote the virtual revolute joints of the robot body for the yaw, pitch, roll in the global frame. $q_7, q_8, q_9, q_{10}, q_{11}, q_{12}$ denote the revolute joints of the manipulator. The underwater robot includes 12 joints to perform tasks, of which there are six virtual joints of the body and six joints for the manipulator. The mass of the underwater robot's body is 136.6 kg and the mass of the manipulator in the water is 31.4 kg.

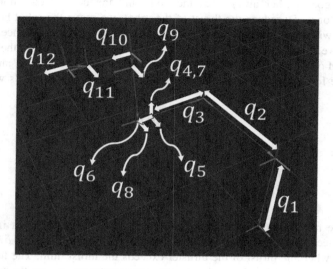

Fig. 2. Kinematics of a 12-DOF underwater robot with the Arm 7E mini manipulator.

2.2 Motion Planning

In this section, we explain how to plan motions with a 12-DOF robot. There are three tasks for controlling the robot, robot body orientation, and end-effector position and orientation. The robot body position is not considered to be a task and remains redundant. The equation for the motion of the underwater robot is given by

$$\delta q = J^+ \delta x + (I - J^+ J)\delta q_0. \tag{1}$$

q denotes the 12×1 vector of the joint angle that includes the virtual joints of the body and the joint of the manipulator. x denotes the 8×1 vector of the position and orientation of the end-effector and orientation of the body, roll, and pitch in the global frame. J denotes the 8×12 Jacobian matrix of the position and orientation of the end-effector and orientation of the robot body. J^+ denotes the pseudo inverse of J. I denotes the 12×12 identity matrix. q_0 denotes the 12×1 vector of sub-task joint angle. The inputs of the robot are x and q_0, and these represent a main task and a sub task respectively. To grasp an object, we control the position and orientation of the end-effector in the global frame, to maintain a stable body state, we control the orientation of the body. While performing a main task, we reduce the thrust of the body using a sub task, q_0.

2.3 Sub Task

When the underwater robot grasps an object, the manipulator reaches out to the object in order to grasp it. The reaching out of manipulator means that the COM of the manipulator goes farther away from the body and the body needs more thrust to compensate for it. To reduce the thrust of the robot, we move the robot body and the manipulator without disturbing the main task using null-space compliance.

At first, we identify the relationship between the limit of the thrust of the robot body and the COM of the manipulator. The 8×1 vector of the thruster force T and the 3×1 vector of the force and the torque on the body due to the manipulator's weight, F is given by

$$-\delta F = B\delta T. \tag{2}$$

where, δT is

$$\delta T = \begin{cases} if\ T_i > T_{i,limit}, & \delta T_i = T_i - T_{i,limit}. \\ if\ T_i < T_{i,limit}, & \delta T_i = 0. \end{cases} \tag{3}$$

i denotes the number of thrusters and B denotes the 3×8 matrix that indicates the relation between the thruster force and the thrust of the body. We define δT to activate the sub task when the thrust of the thruster exceeds the thrust limit. The force and the torque acting on the robot body due to the manipulator mass and the COM of the manipulator is given by

$$m_m \hat{g}^T \delta x_{com} = \delta F. \tag{4}$$

where m_m denotes the mass of the manipulator, x_{com} denotes the 3×1 COM vector of the manipulator F, g denotes the 3×1 vector of gravity and the 3×3 matrix $\hat{g}$ is given as

$$\hat{g} = \begin{bmatrix} 0 & -g_3 & g_2 \\ g_3 & 0 & -g_1 \\ -g_2 & g_1 & 0 \end{bmatrix}. \tag{5}$$

where,

$$g = [g_1, g_2, g_3]^T. \tag{6}$$

By combining Eqs. (2) and (4), we get Eq. (7) as

$$\delta x_{com} = -\frac{1}{m_m} \hat{g}^{-T} B\delta T. \tag{7}$$

The relationship between sub tasks, δq_0 and δx_{com} is

$$\delta q_0 = J^+_{com} \delta x_{com}. \tag{8}$$

where, J_{com} denotes the 3×12 center-of-mass Jacobian matrix of manipulator and J^+_{com} denotes the pseudo inverse matrix of J_{com}. Combining Eqs. (7) and (8), the relationship between the sub task δq_0 and thruster's force, δT in Eq. (9)

$$\delta q_0 = -\frac{1}{m_m} J^+_{com} \hat{g}^{-T} B \delta T. \tag{9}$$

The sub task δq_0 makes the robot body move axially decreasing the thrust of the robot by moving the body and the manipulator without disturbing the main task by multiplying the null-space matrix N. Here N is given by,

$$N = (I - J^+ J) \tag{10}$$

Then the motion q is

$$\delta q = J^+ \delta x - \frac{1}{m_m} N J^+_{com} \hat{g}^{-T} B \delta T. \tag{11}$$

3 Simulation

The simulation environment was developed using Free-floating Gazebo, an underwater environment simulation software that added the functions of thrust and buoyancy.

While the underwater robot carried out the main task, moving the robot body and reaching the end-effector to the point where it was farther away from the initial end-effector position, the thrust of the robot exceeds its limit by activating the sub task. The simulation results are shown in Figs. 3 and 4.

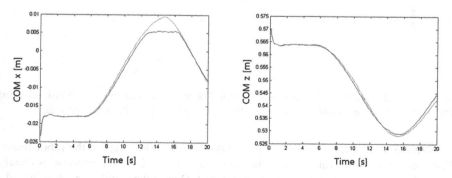

Fig. 3. Shift of the COM of the manipulator in the robot frame.

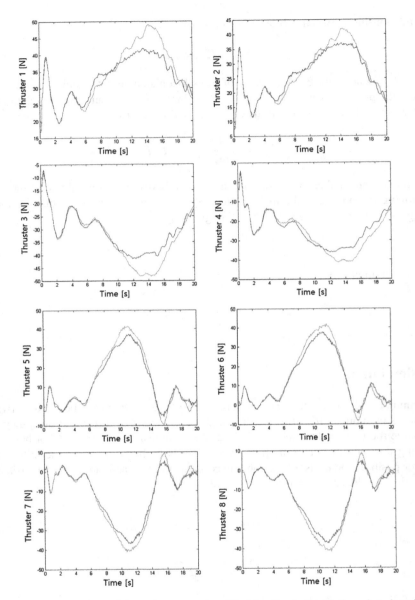

Fig. 4. Eight thrust plots of the underwater robot. The blue line indicates the sub-task activated state, and the red line indicates the non-sub-task activated state. (Color figure online)

Figure 3 shows the transition of the COM of x, z coordinates in the robot frame. When the sub-task activated states, the decrease of the COM of x direction is clearly shown. Figure 4 represents the reduction in thrust. During the simulation when the sub-task is activated, the position of the manipulator COM is closer to the robot body the thrust of the robot is lower than that during non-activated simulation. The thrusters 1, 2, 3, 4 are vertical thrusters used for the vertical movement for the body and compensating for the weight of the manipulator. The thrusters 5, 6, 7, 8 are horizontal thrusters used for horizontal movement for the body. The decrease of the vertical thrusters is specifically shown, because the transition of the torque on the body due to the transition of the COM of manipulator is almost compensated by the virtual thrusters.

4 Conclusion

In this paper, motion planning that reduces the thrust of the robot using null-space is presented. The thrust is reduced by moving the COM of the manipulator by moving the robot's position and the manipulator's joint angle. The relocation of the manipulator's COM does not disturb the main task by multiplying the null-space matrix. The result of simulation shows the shift of the COM of the manipulator and the decrease of the thrust of the robot.

We are going to use motion planning for experimenting with a real robot through teleoperation and autonomous manipulation and add more sub tasks with task-priority redundancy for dexterous and autonomous underwater manipulation.

References

1. Zghal, H., Dubey, R.V., Euler, J.A.: Efficient gradient projection optimization for manipulators with multiple degrees of redundancy. In: 1990 IEEE International Conference on Robotics and Automation, Cincinnati, OH, USA, May 1990, vol. 2, pp. 1006–1011 (1998)
2. Nakamura, Y., Hanafusa, H.: Inverse kinematic solutions with singularity robustness for robot manipulator control. IEEE/ASME J. Dyn. Syst. Meas. Control **108**(3), 163–171 (1986)
3. Antonelli, G., Chiaverini, S.: Task-priority redundancy resolution for underwater vehicle-manipulator systems. In: 1998 IEEE International Conference on Robotics and Automation, Leuven, Belgium, May 1998, CAT. No. 98CH36146 (1998)
4. Kang, J.: Experimental study of dynamic stability of underwater vehicle-manipulator system using zero moment point. J. Mar. Sci. Technol. **25**(6), 767–774 (2017)
5. Kermorgant, O.: A dynamic simulator for underwater vehicle-manipulators. In: International Conference on Simulation, Modeling, and Programming for Autonomous Robots Simpar, Bergamo, Italy, October 2014, HAL-01065812v2 (2014)

Match Outcomes Prediction of Six Top English Premier League Clubs via Machine Learning Technique

Rabiu Muazu Musa, Anwar P. P. Abdul Majeed[✉],
Mohd Azraai Mohd Razman, and Mohd Ali Hanafiah Shaharudin

Innovative Manufacturing, Mechatronics and Sports,
Faculty of Manufacturing Engineering, Universiti Malaysia Pahang,
26600 Pekan, Malaysia
amajeed@ump.edu.my

Abstract. The English Premier League (EPL) is one of the most widely covered league in the world. The prediction of football matches, particularly EPL has received due attention over the past two decades by means of both conventional statistical and machine learning approaches. More often than not, the predictions reported in the literature have rather been dissatisfactory in forecasting the outcome of the matches. This work offers a unique approach in predicting EPL match outcomes, i.e., win, lose or draw by considering top six teams in the league namely Manchester United, Manchester City, Liverpool, Arsenal, Chelsea and Tottenham Hotspur over the span of four consecutive seasons from 2013 to 2016. Fifteen features were selected based on their relevance to the game. Six different Support Vector Machine (SVM) model variations viz. linear, quadratic, cubic, fine radial basis function (RBF), medium RBF, as well as course RBF were developed to predict the match outcomes. A five-fold cross-validation technique was employed whilst, a separate fresh data was supplied to the best model developed in evaluating the predictive efficacy of the model. It was demonstrated from the study that the linear SVM model provided an excellent prediction accuracy of 100% on both the trained as well as untrained data. Therefore, it could be concluded that the selection of the relevant features, as well as the methodology employed, could yield a reliable prediction of top six EPL clubs match outcomes.

Keywords: Support Vector Machine · Football · Match outcome ·
Feature selection

1 Introduction

The forecasting or the prediction of a football match outcome has received considerable attention by the scientific community at large owing to the unique interest in providing meaning to the game play that could improve a team's performance and management. The prediction of the matches often entails the winning, losing or the draw of either the home or away team. Nonetheless, the prediction of the results is somewhat intricate due to the myriad of factors that should be taken into consideration, for instance a team or a

© Springer Nature Singapore Pte Ltd. 2019
J.-H. Kim et al. (Eds.): RiTA 2018, CCIS 1015, pp. 236–244, 2019.
https://doi.org/10.1007/978-981-13-7780-8_20

club may possess all the essential match indicators such as possession, successful pass, good finish, and yet the team is unable to win the match against their opponent [1]. As a result, a number of studies carried out that are available in the literature are unable to demonstrate reliable models with sound prediction [2–4].

The aim of match prediction by means of statistical tools is to outperform the forecasting provided by bookmakers which are solely grounded on subjective and other observational methods that are used to set odds on the results of football matches [5]. The most prominent used statistical means of prediction is ranking. The ranking system is merely a technique of assigning a rank to each club in relation to their previous match results, as such the highest rank is given to the strongest club. The match status can, therefore, be predicted by comparing the opponent's ranks [6]. Although the ranking system has been widely recognised in the prediction of football by many football organising body such as FIFA and World Football Elo Ratings, there are certain noticeable drawbacks to football prediction that are based on the ranking system. Ranks attributed to the clubs do not discriminate between the club attacking as well as defensive strengths. The techniques are combined averages which do not account for skill and pattern changes in football teams. Furthermore, the ranking system is mainly developed to sort various football teams based on their average strength but not to predict the results of football games.

Another system of prediction in football is the rating technique. While the ranking system as previously explained as referring solely to team order, in rating techniques each team is progressively assigned a scale of strength indicator. In addition, the rating can be ascribed not only to a team but also to its other essential performance indicators such as the strength of attacking and defence, a home advantage as well as the skills of a specific player within the team [7]. A typical example of a rating technique is the pirating system that gives measures of a relative superiority between teams based on specific pre-defined performance indicators. Despite the broader acceptance of this system and its reported superiority in terms of profitability against the betting market, some of the measures are reliant upon individual opinions and understanding of the game since player as well as a team assessment or rank in a football game may differ from one person to another. These issues lead researchers in search of other scientifically based methods for prediction. The use of artificial intelligence and other conventional means with the aim of providing an objective evaluation of match prediction by considering a set of performance indicators began to receive attention.

Researchers began an investigation on statistical models for football prediction since the 90s. However, the initial model of match analysis data was earlier proposed by [8]. The analysis was conducted using both Poisson distribution and negative binomial distribution. The author reported that both methods could provide an adequate fit for the prediction of results in football matches. Similarly, a series of ball passing during a match play was analysed by means of binomial distribution, and it was concluded that more goals were scored from a series of higher passing sequences as opposed to lesser passes [9]. Although the accuracy level of the predictions was not adequate from the earlier attempts by researchers, it was reported that indeed football outcomes are to some degree predictable and not merely a matter of coincidence or chance [9].

In recent years, machine learning algorithms have gained due attention in predicting soccer match results. Ulmer and Fernandez [10] evaluated different classifiers namely Baseline, Gaussian Naive Bayes (GNB), Multinomial Naive Bayes, Hidden Markov Model, Linear Support Vector Machine (SVM), Radial Basis Function (RBF) SVM and Random Forest in predicting win, draw or lose of EPL for 10 seasons (2002/2003 to 2011/2012). The features selected were the home team, the away team, the score, the winner, and the number of goals for each team. The findings of the study were rather inconclusive, as none of the models evaluated could achieve reasonably well classification with respect to the evaluated data. This could be due to the randomness and the variation of the tactical and technical capabilities employed by the teams over the span of 10 years.

Artificial Neural Network (ANN) and Logistic Regression (LR) has also been utilised in predicting EPL results [4]. Nine features were selected in their investigation, i.e., home and away goals, Home and away shots, home and away corner, home and away odds, home and away attack strength, home and away players' performance index, home and away managers' performance index, home and away managers' win, home and away streak. The data was extracted from a total of 110 matches played in the 2014/2015 EPL season, in which only 20 sets of matches were used to evaluate the models predictive ability. It was shown that the ANN model is able to yield 85% classification accuracy, whilst the LR provided 93% accuracy. However, it is worth noting that direct comparison of the aforesaid models could not be established as the ANN model classified 3 classes namely win, lose or draw, whilst the LR only classifies win or loss. Therefore, the 93% accuracy attained from the LR might be misleading as one of the class has been eliminated whilst evaluating the classification accuracy.

In an extended study, a Gaussian-based SVM has also been investigated in predicting match results of EPL data [3]. The author employed the same methodology with respect to the selection of features mentioned in [4]. Nonetheless, the prediction accuracy attained was merely 53.3% in classifying correctly win, lose or draw. The author suggested that this form of SVM model is unsuitable for the prediction of the match status. It is worth to note, that the undesirable accuracy level may be caused due to the inadequacy of the data trained.

A polynomial classifier was applied to evaluate its ability in predicting the outcome of football matches in different leagues namely EPL, season 2014/2015; La Liga Primera Division (LLPD), season 2014/2015; and Brazilian League Championships, seasons 2010 (BLC 2010) and 2012 (BLC 2012) [11]. The features selected for the EPL and LLPD in the study are home team goals, away team goals, home team goals shots, away team goals shots, home team shots on target, away team shots on target, home team hit woodwork, away team hit woodwork, home team corner, away team corner, home team foul committed, away team foul committed, home team offside, away team offside, home team yellow card, away team yellow card, home team red card and away team red card. However, 54 features were selected for the BLC. The proposed model was evaluated against a number of machine learning algorithms that are available in WEKA namely, naïve Bayes (NB), decision tree (DT), multi-layer perceptron (MLP), radial basis function (RBF) and support vector machine (SVM). It was shown that the proposed algorithm managed to attain an accuracy of 0.99, 0.99, 0.99 and 0.98, for EPL, LLPD, BLC 2010 and BLC 2012, respectively. Although, high classification accuracy was attained, nonetheless, only cross-validation and sliding window techniques were used to evaluate the model on existing data set (one season), no fresh data was supplied to evaluate the efficacy of the prediction model against fresh data.

Amongst the earlier works of predicting football results using machine learning techniques were carried out by Joseph, Fenton and Neil [12]. The authors focused on the prediction capability of expert Bayesian Network (eBN) against decision tree learner (MC4), NB, Data Driven Bayesian, and a k-nearest neighbour classifier in determining the match outcome for Tottenham Hotspur Football Club for a period of two consecutive seasons, i.e., 1995/1996 and 1996/1997. The selection of the seasons is made based on the availability of the key players during the period selected for the aforesaid study. The features considered are the availability of the key players evaluated at that time, the playing of a particular player playing in midfield, the quality of the opposing team, the quality if the Spurs attack, the quality and performance of the Spurs' team as well as the venue, i.e., home or away. The test and training data were separated across the seasons. It was demonstrated from the study that the eBN provided a more accurate representation of the win, lose or draw for Spurs with an accuracy of 59.21% against the evaluated learners which does not surpass the efficacy of eBN in the study. It is noted whilst the study applied a rather appropriate methodology of evaluating a particular football club, the classification accuracy attained could be further improved through the inclusion of other parameters such as reported in [11] as well as the employment of other machine learning algorithms.

In response to the above-mentioned study, Razali et al. [2] applied BN to predict EPL match results, by considering three seasons (2010–2013). The features selected are akin to [11] with the omission of home and away team hit woodwork, home and away team offside as well as the inclusion of half-time home and away team goals and fulltime home and away team goals. A cross-validation technique, i.e., 10-fold cross-validation was used for the entire data set. The prediction accuracy attained by employing the BN was reported to be 75.09%. It was postulated from the study that the overall prediction accuracy is more superior to that obtained in [12]. However, the claim made by the authors are rather misleading, as the methodology applied in both studies are entirely different.

The present study introduces a different approach to predict EPL matches results by considering the top six clubs namely Manchester United, Manchester City, Liverpool, Arsenal, Chelsea and Tottenham Hotspur in the league over the span of four consecutive seasons from 2013 to 2016. The rationale for the selection of the teams is to eliminate the changes of the teams appearing in the league for a given season. Moreover, the teams selected are incumbent throughout the seasons investigated. This is necessary in order to develop a reliable predictive model in forecasting the match results of the teams evaluated in the present study. This study explores six different variations of SVM models namely linear, quadratic, cubic, fine RBF, medium RBF, and Coarse RBF that has yet been investigated. Furthermore, in this study, we separate the data from each season for independent testing of the best-evaluated SVM algorithms to attest its predictive efficacy.

2 Methods

The data set for the present study contained the results of the top six EPL clubs (Manchester United, Manchester City, Liverpool, Arsenal, Chelsea and Tottenham Hotspur) for four consecutive seasons (2013 to 2016) obtained from http://www. football-data.co.uk/. As each of the 6 teams plays against all the other teams twice per season over the span of four years, the total games evaluated are 456 matches in which the data was randomly separated to 319 and 137 for training and independent testing, respectively. The independent testing here essentially refers to fresh, untrained data used to evaluate the efficacy of the best-predicted model in predicting the classes of the outcome of the matches across all seasons and teams. The selection of the features in the present study is made based on relevant literature in their ability to determine the match outcome results. For each game, our dataset included 15 features namely, the home and away team goals [13, 14], the outcome of the match [15, 16], the home and away team shots as well as shots on target [15, 17], the home and away team fouls [14], the home and away team corners [14, 18], the home and away team yellow cards as well as red cards [14]. The outcomes of the match are defined as the home win (H), draw (D) and away win (A).

In the present investigation, Support Vector Machine (SVM) is utilised. SVM is a supervised learning method that is capable of correctly classifying the multi-class classification problem, i.e. outcome of the match evaluated in the present study, namely A, D, and H through the acquisition of the optimal hyperplane [19]. The attainment is achieved through the identification of the greatest distant between the classes that in turn minimises the risk of misclassification for both the training and validation data set. The general SVM classification function is given by

$$f(x) = \sum_{i=1}^{l} y_i \alpha_i K(x_i, x) + b \tag{1}$$

Whereby $K(x_i, x)$ is the kernel function that is used to measure the training vector (x_i, x). In the present investigation, several kernel functions are evaluated with respect to its effectiveness in correctly classifying the A, D, and H. The kernel functions employed are the linear, quadratic, cubic, fine Radial Basis Function (RBF), medium RBF as well as the coarse RBF. The scale of the fine, medium and coarse RBF is defined by $0.25 * p^{0.5}$, $p^{0.5}$, and $4 * p^{0.5}$, respectively where p is the number of predictors, i.e. fourteen. The sequential minimal optimisation algorithm is used as the solver for the training of the model.

The fivefold cross-validation technique was employed as this form of validation method mitigates the problem that arises from overfitting [20–22]. A total of 319 observations is randomly split into five subgroups, and one of the five subgroups are utilised as the testing data, whilst the remaining four are used as the training data for each iteration. The average performance over all the folds is then computed. Moreover, the remaining data from the overall observation, i.e. 137 was consequently used to evaluate the ability of the classifier models developed to ascertain A, D and H, on fresh data. The classification analysis was carried out by means of MATLAB 2016a (Mathworks Inc., Natick, USA).

The different variation of the SVM investigated in this study are assessed by means of a number of performance metrics namely overall classification accuracy (OA), as well as Cohen's kappa (κ). The OA suggests the overall effectiveness of the tested classifier. The Cohen's kappa statistic, κ provides the insight on the degree of consistency between the observers' reliability, and it ranges between 0 to 1 [23]. The general guideline on evaluating the aforesaid metric is if the coefficient value is 0, it suggests that there is no agreement, 0 to 0.20 as slight, 0.21 to 0.40 as fair, 0.41 to 0.60 as moderate, 0.61 to 0.80 as substantial, 0.81 to 0.99 as almost perfect, whilst 1 indicates perfect agreement. The confusion matrix allows for the apparent observation of the correctly classified and misclassified observations.

3 Results and Discussion

It is evident from the tabulated results (Table 1) that the linear-based SVM model provides an excellent classification of the outcome of the match on the trained data set. The quadratic and cubic based SVM models are able to produce reasonably high classification through the evaluation of all the assessed performance variables. This is followed by the medium RBF based SVM model. However, it is evident that the fine RBF is unsuitable for predicting the correct classification for the outcome of the matches. Through the present investigation, it is apparent that the linear and quadratic based kernel functions may be useful in identifying the outcome of the match based on the predefined match features. The OA and the κ evaluations are also illustrated in Figs. 1 and 2, respectively. Moreover, the best SVM model attained, i.e., linear SVM was tested upon fresh data to evaluate its prediction efficacy and it was observed that no misclassification transpired, suggesting that the developed model is sound in predicting the match outcome for the six selected teams.

Table 1. Model evaluation

		Evaluation metrics
Kernel Functions	OA (%)	κ
Linear	100	1
Quadratic	98.75	0.973
Cubic	94.36	0.876
Fine RBF	62.28	0
Medium RBF	83.70	0.623
Coarse RBF	69.91	0.032

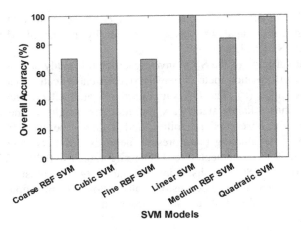

Fig. 1. The overall accuracy of the evaluated SVM models

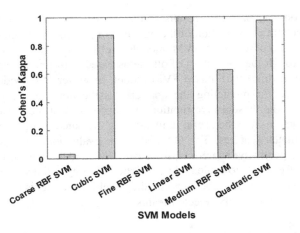

Fig. 2. The Cohen's Kappa evaluation of the trained SVM models.

The findings from the present investigation, suggests that the employment of SVM, particularly linear SVM is able to predict accurately the match status of the six top EPL clubs. Moreover, it could also be concluded, that the selection of the features is appropriate, as it could yield excellent classification predictions. The findings from the study could provide a considerable insight to the coaches with that the selected features are non-trivial in ascertaining the team's probability in losing, winning or draw for a given match. It could be observed from the literature, that the selection of the features adopted by the researchers [2–4, 11, 12], could not yield the classification accuracy as attained in the present study, suggesting that the features evaluated in this study could be valuable in the body of the literature in predicting football matches results particularly EPL and could be extended to other leagues.

Another possible reason for the inability to attain reasonable accuracy by the aforesaid researchers could be due to the methodological design of their study. It is evident that a number of researchers [2–4, 10] neglected or fail to consider the movement (relegation and/or promotion) of the teams between division across the span of the seasons investigated as suggested by [16]. However, it is worth to note, that the present study, adapted the route taken by [12], in which a particular club viz. Spurs is monitored across the seasons evaluated, whilst in this study considered top six EPL clubs. Although the classification accuracy reported by [12] is rather low, the selection of the features as well as the selection of the machine learning algorithms that might not be suitable for the set of data investigated. The present study, took a different approach to mitigate the issues observed in [12], by incorporating different machine learning algorithm extended seasons observation [24, 25], the number of clubs assessed as well as the features selected.

4   Conclusion

The present study investigates the efficacy of a variation of SVM models in predicting the outcome of top six EPL clubs throughout the 2013–2016 season. It was demonstrated from the study that the linear SVM model was able to provide exceptional prediction capability on both trained and untrained data set. The excellent prediction observed could be due to the different methodological approach that was not considered in the existing literature, i.e., the selection of relevant features, the extended span of observation and limiting the observation to the top six teams that are unaffected by relegation and promotion. The adopted unique methodology employed in the present study could be further extended in other leagues as well as different sports.

Acknowledgement. The authors would like to gratefully acknowledge Universiti Malaysia Pahang for funding this study via RDU 180321.

References

1. Brooks, J., Kerr, M., Guttag, J.: Using machine learning to draw inferences from pass location data in soccer. Stat. Anal. Data Min. ASA Data Sci. J. **9**, 338–349 (2016). https://doi.org/10.1002/sam.11318
2. Razali, N., Mustapha, A., Yatim, F.A., Ab Aziz, R.: Predicting football matches results using Bayesian networks for english premier league (EPL). IOP Conf. Ser. Mater. Sci. Eng. **226**, 012099 (2017). https://doi.org/10.1088/1757-899X/226/1/012099
3. Igiri, C.P.: Support vector machine-based prediction system for a football match result. IOSR J. Comput. Eng. **17**, 2278–2661 (2015). https://doi.org/10.9790/0661-17332126
4. Peace, C., Okechukwu, E.: An improved prediction system for football a match result. IOSR J. Eng. **04**, 2250–3021 (2014)
5. Cortis, D.: Expected values and variances in bookmaker payouts: a theoretical approach towards setting limits on odds. J. Predict. Mark. **9**, 1–14 (2015)

6. Min, B., Kim, J., Choe, C., et al.: A compound framework for sports results prediction: a football case study. Knowl.-Based Syst. **21**, 551–562 (2008). https://doi.org/10.1016/j.knosys.2008.03.016

7. Constantinou, A.C., Fenton, N.E.: Determining the level of ability of football teams by dynamic ratings based on the relative discrepancies in scores between adversaries. J. Quant. Anal. Sport **9**. https://doi.org/10.1515/jqas-2012-0036

8. Moroney, M.J.: Facts from Figures, 2nd edn. Penlllin Book Ltd, Harmondsworth (1953)

9. Reep, C., Benjamin, B.: Skill and chance in association football. J. R. Stat. Soc. Ser. A **131**, 581 (1968). https://doi.org/10.2307/2343726

10. Ulmer, B., Fernandez, M.: Predicting Soccer Match Results in the English Premier League. http://cs229.stanford.edu/proj2014/Ben%20Ulmer,%20Matt%20Fernandez,%20Predicting%20Soccer%20Results%20in%20the%20English%20Premier%20League.pdf

11. Martins, R.G., Martins, A.S., Neves, L.A., et al.: Exploring polynomial classifier to predict match results in football championships. Expert Syst. Appl. **83**, 79–93 (2017). https://doi.org/10.1016/J.ESWA.2017.04.040

12. Joseph, A., Fenton, N.E., Neil, M.: Predicting football results using Bayesian nets and other machine learning techniques. Knowl.-Based Syst. **19**, 544–553 (2006). https://doi.org/10.1016/J.KNOSYS.2006.04.011

13. Carmichael, F., Thomas, D.: Home-field effect and team performance. J. Sports Econ. **6**, 264–281 (2005). https://doi.org/10.1177/1527002504266154

14. Palomino, F.A., Rigotti, L., Rustichini, A.: Skill, strategy and passion : an empirical analysis of soccer. Discussion Paper (1998)

15. Miljkovic, D., Gajic, L., Kovacevic, A., Konjovic, Z.: The use of data mining for basketball matches outcomes prediction. In: IEEE 8th International Symposium on Intelligent Systems and Informatics, pp. 309–312. IEEE (2010)

16. Aranda-Corral, G.A., Borrego-Díaz, J., Galán-Páez, J.: Complex concept lattices for simulating human prediction in sport. J. Syst. Sci. Complex. **26**, 117–136 (2013). https://doi.org/10.1007/s11424-013-2288-x

17. Baio, G., Blangiardo, M.: Bayesian hierarchical model for the prediction of football results. J. Appl. Stat. **37**, 253–264 (2010). https://doi.org/10.1080/02664760802684177

18. Tax, N., Joustra, Y.: Predicting the Dutch football competition using public data: a machine learning approach. Trans. Knowl. Data Eng. **10**, 1–13 (2015). https://doi.org/10.13140/RG.2.1.1383.4729

19. Cortes, C., Vapnik, V.: Support-vector networks. Mach. Learn. **20**, 273–297 (1995)

20. Taha, Z., Musa, R.M., Abdul Majeed, A.P.P., et al.: The identification of high potential archers based on fitness and motor ability variables: a support vector machine approach. Hum. Mov. Sci. **57**, 184–193 (2018). https://doi.org/10.1016/j.humov.2017.12.008

21. Akay, M.F., Abut, F., Daneshvar, S., Heil, D.: Prediction of upper body power of cross-country skiers using support vector machines. Arab. J. Sci. Eng. **40**, 1045–1055 (2015). https://doi.org/10.1007/s13369-015-1588-y

22. Muazu Musa, R., Taha, Z., Abdul Majeed, A.P.P., Abdullah, M.R.: Machine Learning in Sports. SAST. Springer, Singapore (2019). https://doi.org/10.1007/978-981-13-2592-2

23. Landis, J.R., Koch, G.G.: The measurement of observer agreement for categorical data. Biometrics **33**, 159–174 (1977)

24. Goddard, J.: Who wins the football? Significance **3**, 16–19 (2006). https://doi.org/10.1111/j.1740-9713.2006.00145.x

25. Heuer, A., Rubner, O.: Fitness, chance, and myths: an objective view on soccer results. Eur. Phys. J. B **67**, 445–458 (2009). https://doi.org/10.1140/epjb/e2009-00024-8

Author Index

Printed in the United States
By Bookmasters

Printed in the United States
By Bookmasters